A BEGINNER'S GUIDE TO 3D MODELING

A BEGINNER'S GUIDE TO 3D MODELING

A Guide to Autodesk Fusion 360

by Cameron Coward

**no starch
press**

San Francisco

Printed in USA

First printing

23 22 21 20 19   1 2 3 4 5 6 7 8 9

ISBN-10: 1-59327-926-4
ISBN-13: 978-1-59327-926-4

Publisher: William Pollock
Production Editor: Janelle Ludowise
Cover Illustration: Limehouse Design
Interior Design: Octopod Studios
Developmental Editors: Frances Saux and Zach Lebowski
Technical Reviewer: Aron Rubin
Copyeditors: Barton D. Reed and Paula L. Fleming
Compositor: Maureen Forys, Happenstance Type-O-Rama
Proofreader: James Fraleigh

The following images are reproduced with permission: Figure 1 was created by DerpyDuckAnimation and is licensed under the Attribution 4.0 International license (*https://commons.wikimedia.org/wiki/File:Tyrannosaurus _rex_3D_render.png*). Figure 1-2 Bundesarchiv, Bild 183-13757-0006 / CC-BY-SA 3.0 (*https://commons.wikimedia. org/wiki/File:Bundesarchiv_Bild_183-13757-0006,_Ingenieure_am_Reissbrett_zeichnend.jpg*). Figure 1-3 was created by Arnold Reinhold and is licensed under the Creative Commons Attribution-Share Alike 3.0 Unported license (*https://commons.wikimedia.org/wiki/File:Computervision_piping.agr.jpg*). Figure 1-4 courtesy of Mark Rogers of *https://www.arclock.design/*. Figure 9-1 was created by Miles Bader and is licensed under the Creative Commons Attribution 2.0 Generic license (*https://commons.wikimedia.org/wiki/File:Tpe10-bloom-f25-l7-1920.jpg*).

For information on distribution, translations, or bulk sales, please contact No Starch Press, Inc. directly:
No Starch Press, Inc.
245 8th Street, San Francisco, CA 94103
phone: 1.415.863.9900; info@nostarch.com
www.nostarch.com

Library of Congress Cataloging-in-Publication Data

Names: Coward, Cameron, author.
Title: A beginner's guide to 3D modeling : a guide to Autodesk Fusion 360 /
   Cameron Coward.
Other titles: Beginner's guide to three dimensional modeling
Description: First edition. | San Francisco : No Starch Press, Inc., [2019]
Identifiers: LCCN 2019003389 (print) | LCCN 2019009343 (ebook) | ISBN
   9781593279271 (epub) | ISBN 1593279272 (epub) | ISBN 9781593279264 (print) |
   ISBN 1593279264 (print)
Subjects: LCSH: Fusion 360--Popular works. | Three-dimensional
   modeling--Popular works. | Computer-aided design--Popular works.
Classification: LCC T386.F87 (ebook) | LCC T386.F87 C69 2019 (print) | DDC
   620/.0042028553--dc23
LC record available at https://lccn.loc.gov/2019003389

This book is dedicated to my mother.
I know she would have been proud.

About the Author

Cameron Coward is an author, maker, and former mechanical designer/drafter. He got his start in CAD while working on a degree in drafting, and spent several years modeling parts and assemblies in the medical, automotive, and furniture industries. These days, he focuses on CAD modeling for 3D printing and other hobby projects. He is a regular contributing author for *Hackster.io* and *Hackaday.com*, and is the author of *Idiot's Guides: 3D Printing*.

About the Technical Reviewer

Aron Rubin is currently Chief Scientist at Dynamic Imaging Systems Inc., where he is developing new products for law enforcement. Prior to that, Aron served as a Senior Software Engineer in Lockheed Martin's Advanced Technology Laboratories, where he helped develop breakthrough solutions in the fields of robotics, machine autonomy, machine vision, and more. Aron's love of CAD started in childhood, when he became fascinated with 3D rendering. He has since applied CAD and CAM to hundreds of projects fabricated through machining, 3D printing, molding, woodworking, circuit printing, and screen rendering.

BRIEF CONTENTS

CONTENTS IN DETAIL

ACKNOWLEDGMENTS

Writing a book, even on a subject you love, is a long and laborious task that can't be done without a great team of people helping you along the way. I found that team at No Starch Press, and would like to thank everyone there who was instrumental in bringing this book to you. Specifically, I want to thank Bill Pollock, Janelle Ludowise, Annie Choi, Zach Lebowski, Frances Saux, Meg Sneeringer, Liz Chadwick, Rhiannon Elliot, Amanda Hariri, Aron Rubin, and everyone else who was working behind the scenes.

An endeavor like this also requires a lot of support from friends, family, and loved ones. I'd like to acknowledge the people who helped me get through the writing process with motivation and support, including Stephanie, Daniel, Sean, Samantha, Dad, Danielle, and Shelby. Thank you all for being there for me!

Finally, I want to thank you, the maker community, and everyone at *Hackster.io* and *Hackaday.com*. Without such a fantastic group of enthusiastic people, there never would have been a market for my writing or this book. The maker community has provided both a resource for me to learn, and an audience for my writing about what I've learned. Most importantly, I have found a community where I feel like I belong.

INTRODUCTION

3D printing technologies have revolution-ized the engineering world. Just a decade or two ago, designers used to spend weeks constructing prototypes that now take only a few hours to complete. Similarly, a company might have easily poured thousands of dollars into a single proto-type part, only to find that it didn't work properly.

Today, engineering teams produce and test parts at a trivial cost. You likely purchased a 3D printer for these reasons. Whether you're a maker, a hacker, an artist, or simply a tinkerer, you wanted the freedom to create your own physical objects.

But while operating a 3D printer is easy, the process of actually design-ing your models has a significantly steeper learning curve. If you're like most people, you've probably downloaded and printed models you found online. That may be entertaining for a time, but you bought a 3D printer, or gained access to one through your school or makerspace, in order to bring your ideas to life, not just to replicate the designs created by others.

In this book, you'll learn everything you need to know to design your very own 3D models with Autodesk Fusion 360, a parametric modeling software that's free for noncommercial use. (If you'd like to take advantage of these skills at work, you can also get a commercial license.) I'll walk you through the fundamentals of the modeling process and then introduce you to advanced tools, drafting, and rendering.

To get started, let's discuss some basic concepts to help you determine whether Autodesk Fusion 360 is the right software to create the models you want.

Who This Book Is For

This book is for anyone who wants to learn how to operate Autodesk Fusion 360, which is parametric computer-aided design (CAD) software you can use to create models for 3D printing or manufacturing. Fusion 360 is easily the most advanced CAD software that's free for hobbyists to use. That said, you won't need any previous experience with CAD or 3D modeling software—or even with 3D printing, if you plan to have your models printed professionally—to work through this book.

What Is Parametric Modeling?

We can categorize today's 3D modeling software into two broad types: *parametric* modeling and *mesh* modeling. Although the two types often overlap, people generally use parametric modeling software for engineering purposes and mesh modeling software for artistic purposes.

In mesh modeling software, you push and pull a virtual mesh to create a 3D model like a sculptor. Digital artists use mesh modeling programs like Blender, Maya, and ZBrush to create video game characters, animations, photorealistic renders, and other forms of three-dimensional art. You can see an example of a digital sculpture made with mesh modeling software in Figure 1.

Precise dimensions matter less in mesh modeling, because people typically use mesh modelers to create sculptures, figurines, and digital assets. In a video game, your character's car doesn't actually have to work; it just needs to look realistic. The game designer shouldn't waste time modeling the thousands of individual parts that make up a working car or ensuring that those parts really fit together.

In contrast, parametric modeling software like SolidWorks, Creo, Inventor, and Fusion 360 (which we'll use in this book) rely on *parameters*, or a shape's dimensions, to create a 3D model. You'd define a simple cube, for example, by providing its origin point in space as well as its height, width, and depth and the units used to describe those dimensions—inches, millimeters, centimeters, and so on. You can then save all those parameters to edit or reference later.

Figure 1: An artistic model designed with mesh modeling software

Why Use Parametric Modeling Software?

If you're reading this book, then you're most likely already interested in parametric modeling. Virtually all modern CAD software is parameter based (though many include some mesh modeling capabilities), and in most cases it's the best option for designing real physical parts. To make parts for 3D printing or CNC milling, you'll need to use software like Fusion 360, especially if those parts will fit together into larger assemblies or mate with an existing object.

Engineers, mechanical designers, and drafters use parametric modeling software because they need to specify the exact size and shape of every feature of a part, and they need to simulate and test those parts. When automotive engineers design a functioning car, they model all of the individual parts, and they're not just modeling them to *look* right; they need them to actually *work* right. Imagine an engineer designing a protective case for your smartphone without specifying its exact size or shape— getting a proper fit would be nearly impossible. Engineers should also be able to easily retrieve geometry and dimension information. If you're having a part manually machined, the machinist needs accurate technical drawings, which you can't directly create within mesh modeling software.

Modeling based on parameters allows you to reference parameters across the different features of a part. For example, you don't need to define a cube's width and depth, because they're always going to match the height, as in Figure 2. All you need to do is set the width and depth as equal to the height.

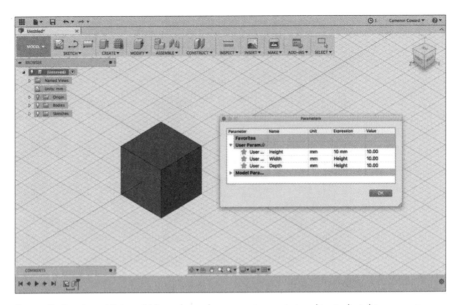

Figure 2: A cube with its width and depth parameters set equal to its height parameter

You can save all these parameters for later use, and the entire model updates automatically whenever they're changed. If you were to go back and change the height of the cube, the width and depth would change, as well, to remain equal to the height.

You can also use parameters to create far more robust relationships between dimensions. For example, you can use formula-driven parameters to change the size of a 3D model of a measuring cup based on a desired volume you specify. With a little bit of a planning and forethought, you can use parameters to create truly sophisticated models.

Finally, you can take advantage of parameters even after you've completed the actual modeling. For example, when drafting a technical drawing, you can use those parameters to add the dimensions of a part to the drawing. You can also calculate useful information, such as the part's volume or mass, based on the model. You can even conduct simulations and tests based on the available physical data using Fusion 360's built-in simulation workspace.

About This Book

In the coming chapters, you'll learn about many of the tools at your disposal. You'll create simple features that serve as the bread and butter of parametric modeling, and then apply advanced, powerful tools. I'll guide you through the individual steps of the modeling projects, and, with practice and experience, you'll learn how to break an intricate part down into a logical series of features. By the end of the book, you should have developed all the skills you need in order to create any kind of 3D model you can conjure in your mind.

Here's a brief overview of what you'll find in this book:

Chapter 1: A Brief History of CAD covers the technological developments that led to the adoption of CAD in various engineering fields and its applications in 3D modeling.

Chapter 2: Parameter, Features, and the Fusion 360 Workspace introduces you to the Fusion 360 interface and the basic principles of parametric modeling.

Chapter 3: Designing Your First Model gets you started with 3D CAD modeling by creating a simple cube.

Chapter 4: Revolving a 2D Sketch into a 3D Object expands your modeling toolset so you can produce more complex geometry.

Chapter 5: Modeling Assemblies teaches you how to create and assemble multipart designs.

Chapter 6: Modeling with Complex Curves covers specialized tools for creating organic shapes.

Chapter 7: Springs, Screws, and Other Advanced Modeling introduces niche tools and how to use them.

Chapter 8: Drafting walks you through turning your 3D models into technical drawings for manufacturing or patenting.

Chapter 9: Rendering teaches you how to create high-quality, presentation-worthy images of your models.

Chapter 10: Capstone Project: Creating a Robot Arm challenges you to design, model, and build your very own robot arm.

That probably sounds like a lot to cover, but we'll be working through incrementally more complex modeling tasks as we go in order to build your skills. The focus will be on hands-on learning, so you can become acquainted with Fusion 360's tools as you use them in real-world exercises. First, flip to Chapter 1 to get a feel for the history of CAD, how that affected today's software, and why we use it.

1

A BRIEF HISTORY OF CAD

Humans have always created their own designs, and in the dark times before the rise of computers, engineers used *technical drawing* to describe three-dimensional objects.

You've probably done some basic technical drawing yourself at one time or another. Maybe you made a sketch for a coffee table you were building, or you drew a simple floor plan of your bathroom to show a contractor where you wanted that new whirlpool tub.

In both these cases, the idea is the same: to illustrate something physical as clearly as possible. It'd be inconvenient to come home to find that your contractor had blocked the bathroom door with your new bathtub because your drawing was ambiguous.

Drafting and the Industrial Revolution

Demands on technical drawing skills exploded in the 18th century, thanks to the Industrial Revolution. The complexity of both machines and product design grew exponentially, and as a result, required more precise instructions for making parts. A rough sketch might have been enough for a carpenter building a table, but steam engines and industrial tools required drawings that were far more detailed.

This necessity gave rise to the formal discipline of *drafting*, the practice of creating technical drawings, like the patent assembly drawing for a steam engine shown in Figure 1-1. Drafters (also called *draftspeople*) began creating engineering drawings that removed all ambiguity about their designs. The single most important part of the drafting profession is the ability to create drawings that can be interpreted in only *one* way. Factories churn out parts by the thousands; imagine having to scrap all of those parts because a machinist misinterpreted the drawing they were working from.

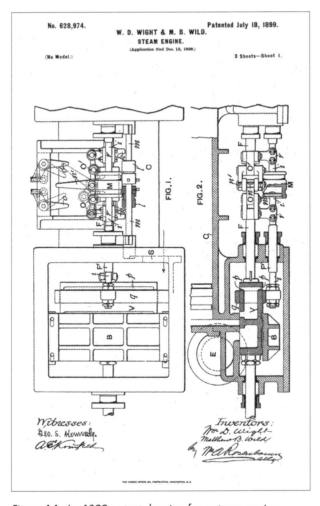

Figure 1-1: An 1899 patent drawing for a steam engine

Of course, ensuring a single interpretation of a drawing requires a common visual language. This design language evolved and became well-defined over time, as the patent system developed and manufacturers began sharing ideas. It eventually standardized through the work of organizations such as the American Society of Mechanical Engineers (ASME), the American National Standards Institute (ANSI), and the International Organization for Standardization (ISO).

While each of these standards differs slightly in its specifics (such as what symbols to use), they all accomplish the same thing: defining a visual language for technical drawings that, when used properly, removes all ambiguity. Someone trained in a specific standard can interpret a drawing created to that standard with total objectivity. These standards encompass all aspects of a technical drawing—from what font a dimension should be to the proper line weights and styles to use for different features of a part. Until recently, drafters hand-drew all of these details using a variety of tools to help do the job, like drafting tables (shown in Figure 1-2), mounted pantographs, and specialized pens, rulers, squares, and curve templates.

Figure 1-2: A large-format drafting table,
with an armature-mounted square

By the latter half of the 20th century, drafters' jobs had become incredibly complex. It wasn't uncommon for a product to have thousands of individual parts, particularly in the automotive and aerospace industries. Manufacturers needed technical drawings for every one of those parts; they also needed drawings illustrating their assembly and even more drawings for tasks like maintenance and repair.

Every single one of those drawings consumed drafters' time, even as the demands of the marketplace led to ever more intricate designs. Luckily, as with just about every other profession on the planet, the burgeoning digital age revolutionized drafting.

Replacing Pen and Paper

Before anyone had even built the first digital computers, manufacturers understood the potential benefits of using computers for engineering. After all, the complex calculations needed for the engineering tasks of the 1940s and 1950s were exactly the kind of problems digital computers would later solve. In 1952, researchers at MIT successfully built a digital *numerical control (NC)* machine too. This became the first milling machine controlled by a computer and laid the groundwork for future *computer-aided design (CAD)* systems.

Although the technology advanced rapidly, the early computers had limited graphics displays, storage capacity, and user interfaces. Nevertheless, the engineering world saw CAD's potential, and any organization with the funds to do so rushed to get in the game. MIT and various automotive and aerospace companies developed proto-CAD systems independently during the '50s and '60s. At MIT, for instance, Ivan Sutherland developed software called SKETCHPAD that allowed an operator to draw with a light pen on a CRT monitor. During a 1963 presentation on SKETCHPAD, Sutherland claimed that "for highly repetitive drawings or drawings where accuracy is required, SKETCHPAD is sufficiently faster than conventional techniques to be worthwhile." (You can read the presentation at *https://www.cl.cam.ac.uk/ techreports/UCAM-CL-TR-574.pdf.*)

Meanwhile, corporations like General Motors (in partnership with IBM), Ford, Citroën, Renault, Lockheed, Boeing, and Bell worked on commercial CAD software. These early CAD systems, like the one shown in Figure 1-3, mostly aimed to replace drafting done by hand, and the ability to create (and modify) technical drawings electronically sped up drafting dramatically.

Figure 1-3: One of the first versions of Computervision CADDS, an early example of CAD software

Entering the Third Dimension

The introduction of affordable personal computers in the '70s led to the widespread adoption of CAD throughout the various engineering disciplines, including mechanical, electrical, and civil engineering.

The new 3D packages that came with these devices changed the design process altogether. Engineers and designers could now create 3D models of large mechanical assemblies, and they could test and verify designs quickly and efficiently, without any physical manufacturing. When it came time to manufacture parts, designers could use the 3D models to quickly generate technical drawings, or even to program machine tools directly, with *computer-aided manufacturing (CAM)* software.

During the '90s, developers refined 3D CAD to take advantage of increased computing power, until most could run on affordable personal computers. Programs of this era include SolidWorks, Solid Edge, and Autodesk Inventor.

CAD in the Modern World

CAD has continued to evolve in the 21st century. In the field of civil engineering, software like Autodesk Revit gives engineers the ability to model entire buildings, down to the electrical system and HVAC layout. Electrical engineers can design printed circuit boards (PCBs) and schematics with software like KiCAD and Autodesk Eagle, which include circuit simulation tools.

This book is about 3D solid modeling from a mechanical engineering perspective, and in that field, CAD has become indispensable. Throughout my career as a mechanical designer, I spent the vast majority of my day-to-day work creating 3D models. Virtually every modern product you come across was first completely modeled by mechanical engineers, designers, and drafters using software.

Today, a drafter makes digital 3D models of parts and assemblies, then uses those models to create technical drawings. Drafters can generate views of the models from any angle and add dimensions, bills of materials, and other technical information very quickly.

CAD for Hobbyists

Until recently, most hobbyists couldn't afford quality CAD software, which was marketed toward professional engineering teams and priced accordingly. Most professional 3D solid–modeling CAD packages cost thousands of dollars, making them impractical for makers and tinkerers working on hobby projects.

That all began to change with the popularity of cheap 3D printers. Suddenly, makers around the world created an immense demand for affordable 3D CAD software. Open source 3D mesh-modelers already existed, but they filled only some of hobbyists' needs; while great for artistic models, they were cumbersome for mechanical tasks.

In the past few years, CAD software developers have started to answer hobbyists' demands for true mechanical, parametric, 3D solid–modeling CAD systems. They started with very basic software, like Autodesk 123D and SketchUp. While these were a step forward (and free), they had serious constraints. Users were mostly limited to working with primitive polygons, and advanced modeling was either very difficult or just plain impossible.

Luckily for the makers and 3D printing enthusiasts of the world, Autodesk released *Fusion 360*, which is what we'll be working with throughout this book. Figure 1-4 shows an example of what is possible with Fusion 360.

Fusion 360 is a fully featured parametric CAD program that has nearly all of the features one would find in costly professional CAD software. When I owned a fabrication business, it was the only CAD program I needed. And, best of all, Fusion 360 is free for noncommercial use.

Figure 1-4: Fusion 360 lets hobbyists create high-quality 3D models for free.

In the following chapters, we'll cover how to use Fusion 360 to create your own professional-quality CAD models. You'll learn everything from basic sketching to advanced modeling techniques. By the end of the book, you'll know everything you need to design complex models for 3D printing and even production manufacturing.

2

PARAMETERS, FEATURES, AND THE FUSION 360 WORKSPACE

The key to understanding how parametric modeling software works is in its name: with parametric modeling, you define every single feature by a collection of characteristics called parameters. For example, a simple cube is constructed by defining variables like its origin, height, width, and depth. Those dimensions are simply the parameters that the software uses to describe the cube. You can then use the software to save all of those parameters and then edit or reference them later.

Using parameter-based modeling is very important in engineering for a few reasons. The most obvious reason is that when you're designing a physical part for the real world, you need to define its dimensions exactly. For example, you can't rely on visual estimates to design a piston that fits into an engine. Instead, you need to explicitly define the piston's exact size and shape through specified dimensions—the parameters that make up your model.

Using Features as Building Blocks

The building blocks of parametric modeling are called *features*. Individual features are often just primitive polygons—basic shapes that you put together with other features to end up with a complex part. One of the fundamental skills you'll learn as a designer is how to break the complex part you envision into a series of basic features.

That process of breaking a component, such as the one shown in Figure 2-1, into individual features is a task that each person will approach differently. There isn't necessarily a right or wrong approach. Two designers might end up with the exact same part using a completely different series of features. Of course, best practices for efficient modeling do exist.

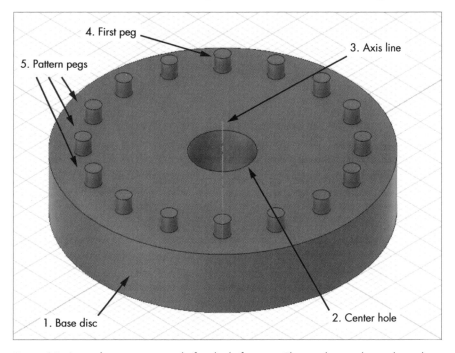

Figure 2-1: A simple part composed of multiple features. The numbers indicate the order the features were added to the part.

If you're a beginner at parametric modeling, you might be tempted to try to use as few features as possible, but that approach generally isn't ideal. Individual features will end up being overly complex, and might contain mistakes that can be hard to find and difficult to modify later. However, the opposite is also true—creating a series of very simple features can be time consuming and inefficient. A balance between complexity and efficiency is best.

Understanding how to find a happy medium is the hallmark of a good designer. With practice and experience, you'll learn how to break down an intricate part into a series of features that follow a logical plan. Which path you choose to follow is up to you as a designer, but the key is to take your time and make a plan.

When you were building a LEGO spaceship when you were little, you probably had an idea of what you wanted the spaceship to look like. But how you actually built that spaceship was completely up to you. Now, let's say your friend came along and decided he wanted to build his own spaceship that looked just like yours. Just because he was able to do so doesn't mean he got there the same way. He could have used a completely different set of bricks and still ended up with a spaceship that had the same final size and shape.

You can think of features in parametric modeling software like those individual LEGO bricks. The bricks come in all different sizes, shapes, and styles, and you can use them for many purposes. In parametric modeling software, one feature can be used to create a basic cube, while another might be used to cut a hole through that cube. Still another could be used to round the edges and corners.

Every feature you'll work with is defined by at least one parameter, such as the depth of a cut. Most features will have numerous parameters that either you set explicitly or the software generates for later use. On many features, like the Extrude shown in Figure 2-2, you'll use parameters in conjunction with *sketches*, which are 2D drawings that are used to make 3D shapes or to define locations or paths.

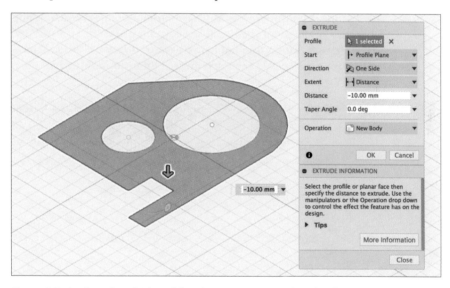

Figure 2-2: An Extrude, which is defined in conjunction with a sketch

You can use a sketch to draw the profile shape of a cavity you're cutting into a part, or to make a path that a piece of tubing will follow. Not all features require a sketch, but most do, and you'll find yourself spending the majority of your modeling time creating sketches.

The Fusion 360 Workspace

Now that you're familiar with the principles of parametric modeling, it's time to get acquainted with the Autodesk Fusion 360 workspace. One of the major factors that holds hobbyists back from jumping into computer-aided design (CAD) is the complexity of the software. The interface is chock full of buttons and toolbars that seem meaningless at first glance, so the uninitiated might feel overwhelmed. CAD becomes significantly less daunting once you grasp the fundamentals of the workspace.

Downloading Fusion 360

Autodesk Fusion 360 is available for Windows and macOS. You can find a download link at *https://www.autodesk.com/products/fusion-360/free-trial/*. Fusion 360 is free for hobbyists and enthusiasts; you only need to pay for it if you're using it for commercial purposes making more than $100,000 per year. You can enter your own name as the company name when asked for it.

Because Fusion 360 is cloud-based software, you need to have a free Autodesk account to access it. Follow the links to download the installer to your computer. Once the software installation is complete, follow the prompts to create an Autodesk account, or sign in if you already have one. If you're using Fusion 360 for noncommercial purposes, be sure to choose the option for hobbyists.

After creating an account and signing in, you'll be asked to choose the units you want to work with. For the sake of easy math, I'll be using millimeters throughout the book, but you can use inches if you prefer. You can always change this setting later or set it for individual files. After you choose the units, you should see a screen that looks something like the one in Figure 2-3.

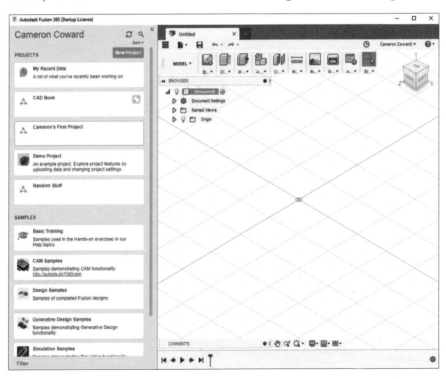

Figure 2-3: The Fusion 360 startup screen

The screen is divided into two panes: the left pane is your project browser, where you can find files (though they're stored in the cloud), and the right pane is your workspace.

Using the Project Browser

By default, you get a *project* called "[your name]'s First Project" where your files will reside. You can use this project or, if you like, create a new one. This project is where your 3D models, drawings, and other files are saved as Fusion 360 documents.

Later, when you're working with larger projects, you can create folders to keep your projects organized. If you collaborate with another user, you could share a project with them to give them access to your files. To open or close the Project Browser (to save monitor real estate), use the icons in Figure 2-4.

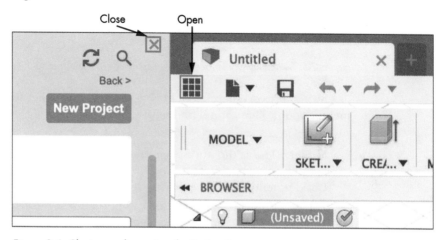

Figure 2-4: Closing and opening the Project Browser

You should only need to have the Project Browser open when you're creating a new project, changing projects, or opening an existing file. Otherwise, you can keep it closed to give yourself as much screen space as possible for the workspace pane, which is where you'll actually be working. The file icon between the Project Browser and Save buttons lets you access menu options for creating a new file, saving a file, exporting a model, and so on.

Switching Workspaces

In the primary workspace pane, you should see the main toolbar across the top of the window. On the left-hand side is a drop-down menu that should be set to "Model" by default. This drop-down menu allows you to switch between workspaces that are set up for different tasks (see Figure 2-5).

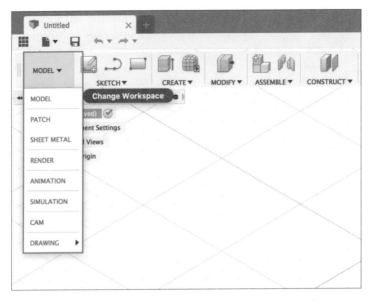

Figure 2-5: Switching workspaces for access to menus and toolbars

Switching workspaces should give you access to other toolbars and menu options. Here, you'll find a list of workspaces along with a brief summary:

Model   This is the workspace you'll probably use the most. It contains all of the *solid modeling* tools, which means that everything you model could potentially exist as a real physical object.

Patch   This workspace contains all of the *surface modeling* features. You'll learn a bit about surface modeling in Chapter 7, but this book will mostly focus on solid modeling.

Render   This workspace contains tools for creating high-quality renders of your 3D models, typically for presentation purposes. Chapter 9 covers the basics of creating renders.

Animation   This workspace lets you animate assemblies in order to illustrate how the movement of a machine (for example) will work. Although this book doesn't cover animation, I suggest you play around in this workspace.

Simulation   You can use this workspace for strength testing and stress-analysis simulations, which are also outside the scope of this book.

CAM   Computer-aided manufacturing (CAM) is one of the major selling points of Autodesk Fusion 360. It allows you to create toolpaths and program various *computer numerical control* (*CNC*) machines. CAM is a complex topic, so it isn't covered in this book.

Drawing   Here, you can create technical drawings that you can use for manufacturing, filing patents, or simply illustrating your model. You'll learn more about this workspace and topic in Chapter 8.

Navigating Around the Model Workspace

The Model workspace is where you'll do most of your work. Navigate to that workspace and take a look around the screen, which should be divided into several key areas, as shown in Figure 2-6.

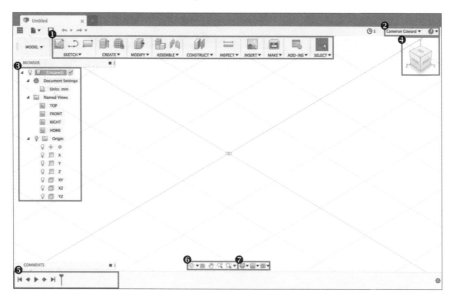

Figure 2-6: The Model workspace

1. The *Ribbon* is the main toolbar where you'll find the sketches and features for creating solid models, along with a few other miscellaneous tools. Generally, you'll work left to right—starting with Sketch, moving to Create, then Modify, and so on.

2. Your Autodesk account options, help, and Fusion 360 preferences and options are accessible via the dropdown menu.

3. The *Component Browser* shows you the views, planes, and other construction geometry you've created—some are created by default. For now, you'll just be working with one component at a time, but if there were more than one, you'd select them here.

4. The *View Cube* allows you to rotate your 3D model in space and quickly orient the viewport to face a particular side of the model.

5. The *Design History Timeline* updates as you create sketches, features, and construction geometry. This maintains a history of everything you do and allows you to go back and make changes as needed.

6. Additional view tools for orbiting your model, fitting it to the window, panning, and zooming.

7. Various display settings and options. You can use this toolbar to change the look and feel of the workspace and model, to set up the view grid and determine how it acts, and to divide the view window into multiple viewports.

The *viewport* is where you'll be actually interacting with your 3D models (see Figure 2-7). This is an essentially infinite space where the model will exist. Because you haven't created a model yet, you should see a grid laid out over a plane composed of two axes—in this case, the x-axis and z-axis—and centered on the *origin*, which is the point where the x-axis, y-axis, and z-axis all meet.

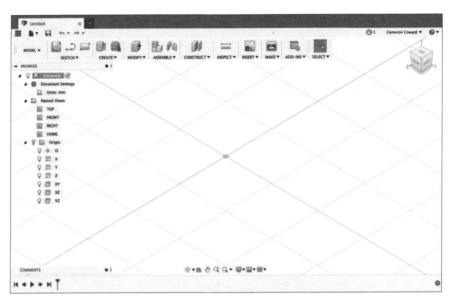

Figure 2-7: The viewport is where you will interact with your 3D model.

Use the View Cube to rotate the viewport around the origin and a model, or use the orbit tool from the Navigation toolbar to freely orbit or rotate the model. Holding down the scroll wheel of your mouse pans the toolbar. Zoom in and out with the magnifying-glass buttons at the bottom of the window, or pan the viewport by selecting the hand icon from the View toolbar. You can also pan with your mouse wheel or two fingers on laptop touchpads.

Understanding the Model Workspace's Main Toolbar

Each workspace has its own main toolbar, and the Model workspace's toolbar contains everything you need for solid modeling. A 2D plane can only exist as either a surface or as *reference geometry*, which means it isn't part of the solid model. This is because a 2D model has no thickness, so it isn't physically possible to create in the real world. Working with solid models ensures that the parts you create are (theoretically, at least) actually makeable. This is in contrast to surface modeling, where you have the ability to create surfaces that have zero thickness, which is physically impossible.

In the following chapters, you'll learn how to use each tool in the Model workspace. But for now, here is a brief overview of the drop-down menus in the main toolbar (Figure 2-8):

Figure 2-8: The Model workspace's main toolbar

Sketch   The various tools for creating a sketch and drawing within the sketches you create.

Create   This is where you'll find the tools for creating solids. Some of these require a sketch, but there are also features for creating basic primitives without the need for a sketch.

Modify   Features for modifying an existing solid, such as adding a fillet to round the edges of a 3D model. This drop-down also has options for setting the *material* and *appearance* of your model. Setting the material allows you to calculate things like the mass and to perform simulations. Appearance just affects how the model looks, which is useful when creating renders. This is also where you'll find your custom parameters.

Assemble   The tools for putting together multiple components to form an assembly. This is useful for checking to make sure your parts all fit together, testing their motion, and creating assembly drawings.

Construct   All of the *construction geometry*, including nonsolid planes, lines, and points that are used to assist in modeling, but which have no actual three-dimensional substance themselves.

Inspect   This is where you'll find a host of tools you can use to analyze your model. Most commonly, you'll use the Measure tool to perform basic tasks like checking the distance from one point to another, but there are also tools for checking the center of gravity, drafting (for molding), and a lot more.

Insert   Use this to bring external resources into your model. For instance, when creating high-quality renders, you might want to insert a *decal*—an image file—for something like the printed face of a dial. You can also pull in external DXF and SVG files (two-dimensional line drawings), and 3D mesh files, like STL files, which you may use for 3D printing.

Make   Mostly used for 3D printing, this drop-down menu has an option to export an STL file. It also has utilities for uploading your model to various paid 3D printing services.

Add-Ins   There are a number of first- and third-party add-ins available for Fusion 360 that give you tools for specialized tasks. You can find them here.

Select   Most of the time, the default selection options work just fine, but once your sketches, models, and assemblies get complex, it can sometimes be difficult to actually click the thing you're trying to select. This drop-down menu gives you options for only selecting particular parts of the sketch or geometry in those situations.

Summary

Now that you have a basic understanding of the interface and workspace, feel free to play around with the different options and settings. If you feel like you've accidentally messed up your settings, you can always revert to the defaults by clicking the username in the top right-hand corner and then clicking **Preferences ▸ Restore Defaults**.

3

DESIGNING YOUR FIRST MODEL

It's finally time to get modeling! Over the next few chapters, you'll gradually learn increasingly complex sketching tools, features, and techniques. By the end of the book, you'll be a master of 3D modeling and have the skills to design whatever you can think of. First, let's start with the basics.

Getting to Know Your Tools

In this chapter, you'll learn to use fundamental sketch tools and features in Fusion 360, which you'll use more than any of the complex tools in the chapters to come. Just like how a skilled carpenter can do amazing things with a handful of basic tools, you'll learn how to construct models using use simple tools like Extrude and Revolve. These foundational tools may not be glamorous or designed for specialized work, but they're powerful and useful for a wide variety of tasks.

Using Basic Sketch Tools

You can find the basic tools under the Sketch drop-down menu, as shown in Figure 3-1. If you don't see the Sketch drop-down menu, make sure you're in the Model workspace.

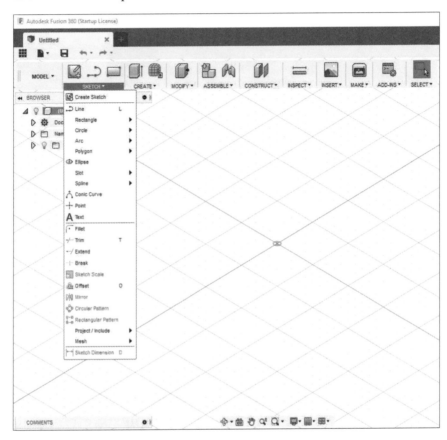

Figure 3-1: The Sketch tools menu

You should see quite a few options here, and many of them have additional submenus. In this chapter, you'll use the following tools:

Create Sketch   Creates a new sketch. When using this tool, you must select on which plane the sketch will be drawn. When you haven't yet started a model, you will usually select one of the x-y, y-z, or z-x planes, and if you've already started building a model, you'll usually select a flat face as a sketch plane. You can also start a sketch by selecting one of the sketch tools, like the Line tool, and then choosing a plane.

Line   Draws straight line segments. You can connect multiple segments together to form shapes. This tool is so frequently used it gets a default keyboard shortcut key: L.

Rectangle   Creates four line segments automatically joined with *constraints*, or rules that specify how sketch entities are linked. For example, a constraint might lock the endpoints of two line segments together, make an arc tangent to a straight line, or make two lines parallel. In this case, the Rectangle tool automatically constrains the endpoints of the four lines and makes them perpendicular to each other. The submenu contains options for defining the rectangle, and the keyboard shortcut R draws two-point rectangles that are defined by any two corners.

Circle   Draws circles. Multiple options are available under the submenu for defining and positioning the circle. Sketch circles are true mathematical circles—they're not divided into arcs or a series of line segments. The C shortcut creates a circle defined by its centerpoint and diameter.

Arc   Draws arcs. The submenu provides a few options here for how you want to define the constraints of the arc. Since arcs aren't as common as the other shapes, the Arc tool doesn't have a keyboard shortcut.

Fillet   A *fillet* (pronounced "fill-it") is a frequently-used tool for rounding a sketch's corner. You define it by the radius of the arc that joins the two lines. When you use it in a sketch, this tool automatically cuts two line segments short, places an arc connecting them, and constrains that arc to be tangent to both line segments.

Trim   When two lines intersect, this tool trims one of the lines down to the point where the intersection occurs. The keyboard shortcut is T.

Extend   Extends a line until the point where it intersects with another line. If you apply this tool and there are no other lines to intersect, then nothing will happen—lines can't go on infinitely in Fusion 360.

Offset   Creates a line, or chain of lines, that is offset from the selected line by a defined distance. The keyboard shortcut is O.

Sketch Dimension   Defines the dimensions of your sketches, like the length of a line or the angle between two lines. The keyboard shortcut is D.

The sketch entities you create with these tools will likely form the bulk of your toolset—you can sketch almost anything with just these tools. But you're not reading this book just to make sketches—you want to create 3D models! That's where features come in.

Using Initial Features

You can use the tools in the Create menu (shown in Figure 3-2) to make initial features, which you'll combine to create 3D shapes. You must always use one of these initial features to create your first 3D entity because they don't require any existing model geometry to work off of (though a sketch may be required). This contrasts with the tools from the Modify menu, which only work with existing geometry. Keep in mind that you can use these tools to create features throughout the modeling process, and not just to create initial features.

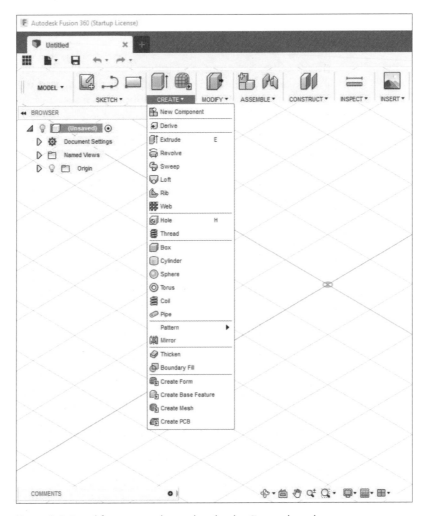

Figure 3-2: Initial features are located under the Create drop-down menu.

Some of these initial features are more frequently used than others. The following list briefly describes the most common features:

Extrude   Adds thickness to a sketch to form a 3D solid. Conversely, an extrude feature is also used to cut into an existing solid in the shape of a selected sketch. This is the most common feature in 3D modeling, and the keyboard shortcut is E.

Revolve   Similar to Extrude, this feature creates a new solid or cuts an existing solid from a sketch. The Revolve feature makes a solid body by revolving the sketch around a selected center axis by a specified number of degrees. For example, a square that touches the axis of revolution would form a cylinder if it were revolved a full 360 degrees.

Hole   Streamlines the process of cutting a hole in a solid. With Extrude and Revolve, you can create a hole of any shape you can sketch, but using the Hole feature can sometimes be faster for creating simple holes. But since cutting holes with Extrude or Revolve is more common and versatile, I typically avoid using Hole.

Box, Cylinder, Sphere, Torus, Coil, and Pipe   These features create primitive solids. Like the Hole feature, these features can all save time in some situations—mostly as basic starting points—but for learning purposes, you won't be using them in this book. You can create each of these (with the possible exception of the Coil feature) with Extrude or Revolve.

You might be surprised that you'll only use two of these initial features for now (Extrude and Revolve), but they're powerful and useful. In fact, Extrude and Revolve alone will probably make up more than 90 percent of the initial features you use in your 3D modeling endeavors.

Modifying Features

While you can create initial features without any existing solids, you'll need to have a 3D object to modify features. Modifying tools are often used toward the end of the modeling process to put the finishing touches on a part you've designed, but you can certainly use them earlier. You can access these tools from the Modify drop-down menu (see Figure 3-3).

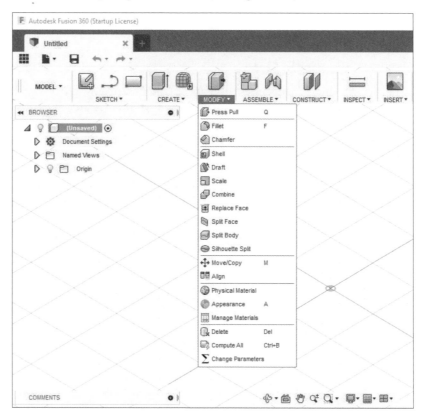

Figure 3-3: You can find modifying features under the Modify drop-down menu.

Like the initial features, you'll use some of these tools more than others, so the following list covers just a few of the available modifying tools in Autodesk Fusion 360:

Fillet   Like the Fillet sketch tool, you can use the Fillet modifying tool to create rounded edges and corners three-dimensionally. You can select inside or outside edges and define a radius. The keyboard shortcut is F.

Chamfer   The Chamfer tool works much in the same way as Fillet, except that it gives the edge a bevel instead of rounding it. You can define a chamfer by the angle of the cut and one distance from an edge, or with the distances from two edges.

Shell   A surprisingly useful feature, the Shell tool hollows out a solid and removes the selected face for an opening. You also need to specify the thickness of the walls, so the model remains a solid. Shell could, for instance, turn a cylinder into a cup or a cube into a box.

Now that you know what's in your tool belt, let's jump into the modeling!

Modeling a Cube

Creating a cube is like the "Hello, World!" of 3D modeling. Although you can quickly and easily create a cube with the Box initial feature, you'll create a cube the hard way from sketches so you can learn about the modeling process.

Creating the Sketch

To model the cube using an Extrude feature, you first need to start with a sketch. Go to the **Sketch** drop-down menu and select the **Create Sketch** option. You should be asked to choose a plane on which to draw the sketch. Figure 3-4 shows the sketch on the xy-plane. The plane you choose is completely up to you—you can always rotate models at a later point, so it's really about how you want to picture the part in your mind.

Once you choose your sketch plane, the viewport should rotate to face that plane. You can rotate the view if you like, but most of the time you'll want to sketch while facing the sketch plane.

Because you're modeling a cube, start by sketching a rectangle. In this case, you want a *center rectangle*. This means the rectangle will be constrained to stay centered on the origin point, or (0, 0, 0). Constraints are geometric rules that apply to sketch entities. For instance, a perpendicular constraint forces two lines to meet at a 90 degree angle. Keeping your models constrained to the origin is typically good practice, because it will make using the default reference planes and axes easier in the future.

Now navigate to **Sketch ▸ Rectangle ▸ Center Rectangle**. A dialog offering various options should pop up on your screen. Clicking the plane determines the centerpoint, so click the origin to automatically create the rectangle's first constraint. Your next click should place a corner of the rectangle, so just click off in space a bit away from the origin so you get something rectangular, as shown in Figure 3-5. The point you click should be mirrored across both the x- and y-axes—automatically placing all four points of the rectangle.

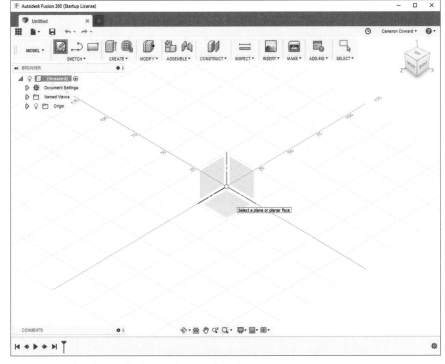

Figure 3-4: Sketch on whichever plane makes the most sense to you.

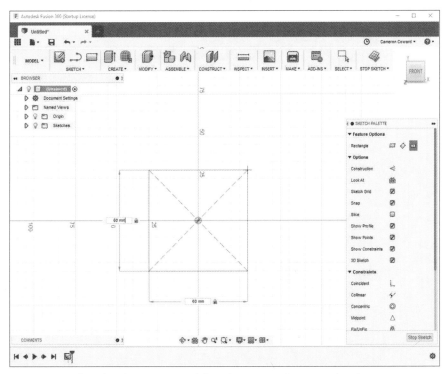

Figure 3-5: When sketching, you can define dimensions during or after the operation.

You may have noticed that, after your second click that set the corner of the rectangle, the dimensions popped up for the width and height of the rectangle. You could have specified one of them, pressed TAB, entered dimensions for the other, and then pressed ENTER. This would have defined the dimensions of the rectangle and constrained it to them.

But because you didn't do that and just clicked a point in space, Fusion drew the rectangle without any associated dimensions. This means that the rectangle is constrained to be centered on the origin, the endpoints of the four lines are automatically locked together with coincident constraints, and the four lines are constrained to be perpendicular to each other. But, it doesn't have any dimensions to constrain its actual size.

Completely Constraining the Sketch

At this point, the four lines of your rectangle should be blue, which by default signifies that a sketch isn't completely constrained. You should never finish a sketch without it being completely constrained with geometrical constraints or dimensions. Until the sketch is completely constrained, points and lines could unintentionally shift, which leads to ambiguity about the accuracy of the part; and when it comes to CAD, ambiguity is always bad.

To completely constrain the rectangle, you need to define the dimensions of the part. Use the D shortcut to create a sketch dimension—or choose it from the Sketch menu—and select a vertical line. Move your cursor off to the side of the line and click to place the dimension. Then enter **50** and press ENTER. Do the same for one of the horizontal lines. The dimensions should display as shown in Figure 3-6.

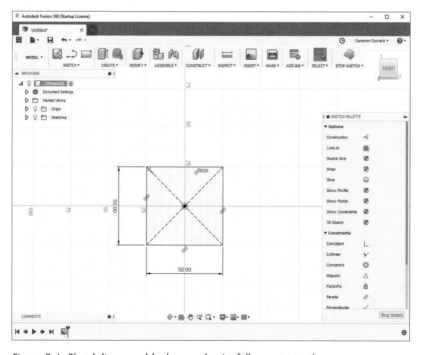

Figure 3-6: Sketch lines are black once they're fully constrained.

Now that you've defined the dimensions, all the lines should turn black in the default Photobooth visual environment (colors may look different if you change the environment). That means you've completely constrained the rectangle, and nothing unexpected should happen later on that might affect the precision of the sketch. Ensuring sketches are always fully constrained should keep you from having to chase down problems when you're working with more complex models. Leaving a sketch partially unconstrained won't break anything, but it can cause unpredictable results if the sketch or features shift.

Extruding the Sketch

To extrude your completely constrained sketch to form a 3D solid, press E or choose **Create ▶ Extrude**. The view should shift to make it easier to tell what your extrusion is doing, and you'll be asked to choose a *profile*, which is any closed loop in a sketch. In this case, your profile is your rectangle, so click somewhere inside it. Like in Figure 3-7, the selected profile should be highlighted. Although it doesn't apply here, you can select multiple profiles if they're on the same plane by clicking inside them when the "profile" field in the dialog is active.

Like all features, Extrude provides you with some options to work with. The Start option is where you want the *zero point*, or the start of the extrusion, to be. In this case, you should stick with the default, which is to start it at the sketch plane.

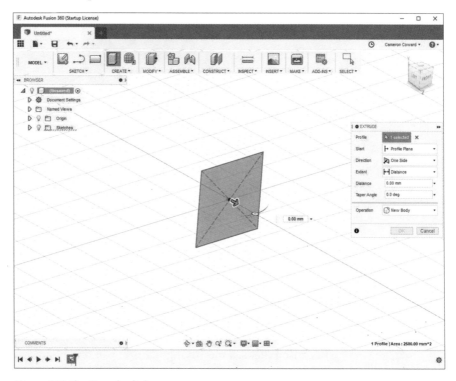

Figure 3-7: The Extrude dialog

The Direction option determines whether the extrusion should extend to one side of the sketch plane, both sides with different distances, or both sides with symmetrical distances. You want your cube to stay centered on the origin point, so choose **Symmetric**. When you select Symmetric, you'll be asked if the dimension refers to total extrusion length or per side. You want total so that all dimensions match. Set the Distance option to 50 mm with a 0 degree taper angle.

The last option, Operation, determines what the feature does. Because you don't have any other solid bodies in your model, New Body is the only option that makes sense, but if you wanted to cut a hole in an existing body, you could use Cut. Other options are Intersect, which keeps just the section where two solids overlap, and Join, which merges two solids together.

Finally, click **OK** to create your very first 3D model! The finished model should look like the cube in Figure 3-8.

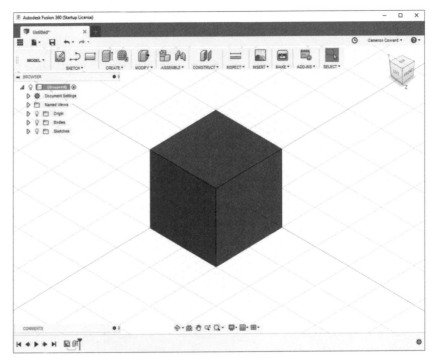

Figure 3-8: A cube, your first 3D model!

Now that you've completed the cube, let's give it some more character with some modifying tools.

Modifying the Cube

In this section, you'll fillet all of the cubes' edges and cut a circular hole into one of its sides. Filleting the edges gives your cube a nice rounded look.

Because Fillet is a modifying feature, you don't have to sketch anything; you only need to select the edges you want to fillet. Be sure to rotate the model to get the back edges and choose a radius of 5 mm, as shown in Figure 3-9.

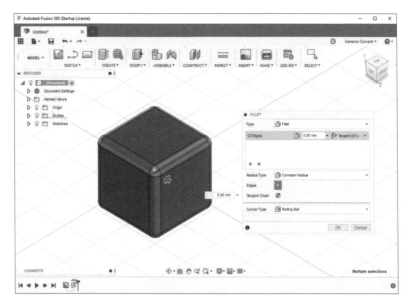

Figure 3-9: Select edges to fillet.

To better understand how the Fillet tool works, try tinkering with the radius measurement and note its behavior.

To cut a hole through one side of the cube, create a new sketch. For the sketch plane, choose either the front or back flat face of the cube (you'll see why soon). Next, sketch a **Center Diameter Circle**, with the center on the origin point and a diameter of 25 mm, as shown in Figure 3-10.

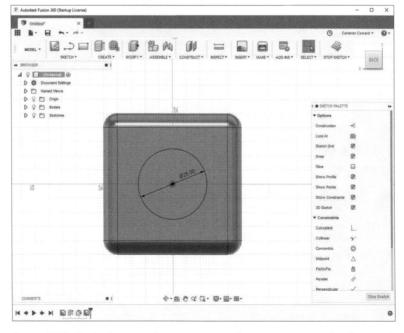

Figure 3-10: Circle drawn with centerpoint on the origin on one side of your cube

This is why you took the time to center your rectangle around the origin and use a symmetric extrusion. When you select a sketch plane, the origin point is automatically centered on your part, because the previous features were modeled symmetrically around the origin. Now when you center your circle on the origin, you know that it's also centered on the face of the cube. If your cube wasn't centered on the origin, you'd have to use additional constraints, construction geometry, or dimensions to center the circle on the face.

With the circle drawn and fully constrained, you can extrude it to cut the hole. To execute a one-sided Extrude, set the Extent option to **All** so that it cuts through all existing solids. You may need to select the **Flip** toggle to change the direction, and then set the operation to **Cut** (see Figure 3-11).

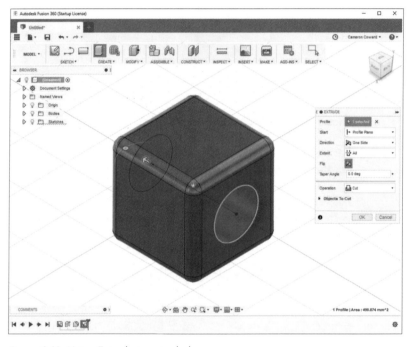

Figure 3-11: Using Extrude to cut a hole

Your cube now has a hole in it—nifty! To save your model, click the **Save** icon.

Using the Design History Timeline

You may have noticed that icons have started appearing at the bottom left of the window in the Design History Timeline. These icons represent each step you've taken to create this particular model (see Figure 3-12).

You can also use the Back and Forward buttons to skip back and forth through all of the features to see each step. You can even use the Play button to see a sort of time lapse of your modeling process.

Being able to skip through the features is cool, but the true power of the Design History Timeline lies in its ability to modify those old features and sketches. When you use the timeline to make changes to an earlier step, all subsequent items in the Timeline should update to reflect your changes. Let's give it a try!

Figure 3-12: The Design History Timeline

Right-click the second item in the Timeline—the first should be a sketch, and the second should be the initial extrusion. Next, click **Edit Feature**, and the original Extrude feature dialog should pop back up. Now change the distance from 50 mm to 100 mm and then click **OK**. The result should look like the model shown in Figure 3-13.

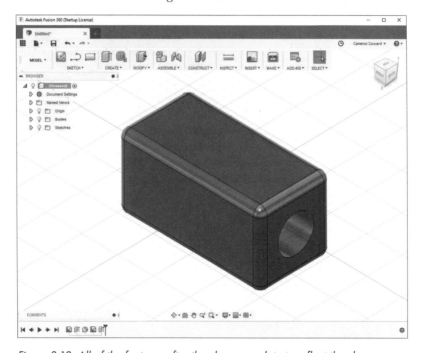

Figure 3-13: All of the features after the change update to reflect the change.

Notice how the edge fillets were lengthened, but the hole still goes all the way through the part. Thanks to the power of parametric modeling, Fusion 360 groups geometry into variable-driven features that adapt to changes. The edge itself had the fillet, so when the edge got longer, the fillet did too. Recall that when you extruded the hole cut, you chose the All option, so when the extrusion was lengthened, so was the cut.

The reason you chose the front or back face for this feature was to avoid an unexpected result, like the one shown in Figure 3-14.

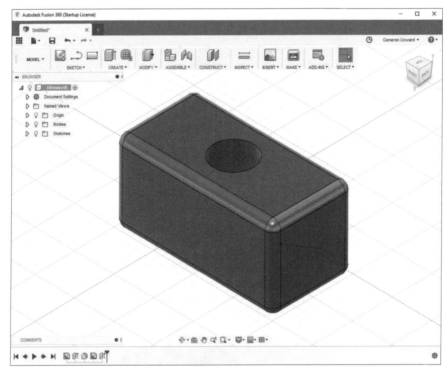

Figure 3-14: The hole is on the wrong face!

In this case, the hole was extruded through one of the side faces. When the model was just a cube, it looked the same no matter which face you sketched the hole on. But when the model was extended, the hole could end up in the wrong place if you chose the left, right, top, or bottom faces. How you choose to orient models is completely up to you, as long as you keep in mind how the features are positioned relative to each other.

Exercises

Practice the techniques you learned in this chapter by completing the following exercises on your own.

Add a Slot

Remove the hole in the block by right-clicking the corresponding feature in the Design History Timeline and selecting **Delete**. In this case, the hole feature is the last one in the Timeline. But, if you ever lose track of the features, you can hover your cursor over the Design History Timeline entry icons to highlight the corresponding feature in the model, or drag the marker back through the Timeline to see each modeling step.

Now you'll add a slot to the top face. To sketch a slot, you can use the Slot tool in the Sketch menu. There are different options for how to define it, and you'll want to use overall length. Give it a length of 50 mm and a width of 20 mm. When you're done, it should look like the model shown in Figure 3-15.

Figure 3-15: Add a slot to the top face of the block.

Make the Slot Responsive

As you've already seen, it's important to consider how models will change in the future. In the first exercise, you made the slot 50 mm long—half the length of the block itself. Suppose you *always* want the slot to be half the block's length, no matter what that length is.

To do that, edit the slot sketch. Under the Modify drop-down, click **Change Parameters**, and a dialog will pop up that lists all of dimensions used in the model (you may have to expand the submenus). Parameters are divided into User Parameters, which you explicitly create, and Model Parameters, which Fusion 360 generates automatically when you model new features. One of the dimensions will be the Extrude Distance of 100 mm. Take note of its name, which was "d3" in my case.

Next, you can modify the slot length dimension with that name so it says something like "d3 / 2". Now, the slot length will always be calculated as being half of the extrude distance.

If you elongate the block, your model should look something like the model shown in Figure 3-16.

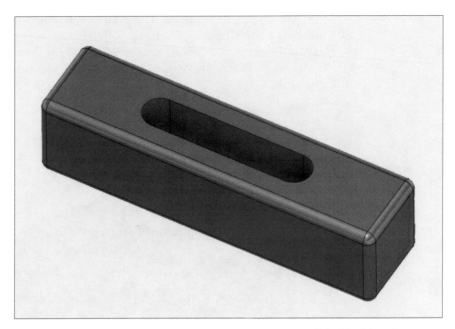

Figure 3-16: Use parameters to make the slot length always half the block length.

Summary

Using a simple example of CAD modeling, you learned how to use Fusion 360's most versatile tool: Extrude. In the following chapters, you'll dive into increasingly complex features, but you'll continue to use Extrude throughout this book and in all of your 3D modeling projects.

4

REVOLVING A 2D SKETCH INTO A 3D OBJECT

The most direct way to produce parametric 3D objects is through manipulating 2D sketches. In the last chapter, you produced a cube by extruding a square. In this chapter, you'll learn how to use the Revolve feature to produce a spherical object from a sketch, then practice tying features together by smoothing them out with fillets and chamfers.

Creating a Sphere as a Revolve Feature

In this section, you'll use the Revolve tool to create a solid body by spinning a profile around a central axis.

Sketching the Circle

Start by creating a sketch on the Front plane; remember to click **Front** on the view cube to orient the view toward the correct plane. All revolves require a central *axis*—an imaginary line around which the sketched geometry is revolved—in order to create a solid. In this case, that central axis will be the existing y-axis.

Draw a Center Diameter Circle with its centerpoint on the origin and a diameter of 50 mm. Then draw a line from the top of the circle to the bottom. Make sure the line is vertical; if it is, your cursor should automatically snap onto the circle. Your sketch should look like Figure 4-1.

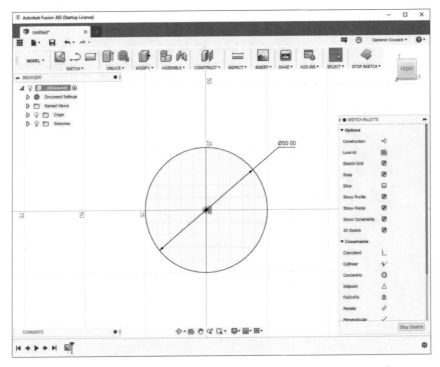

Figure 4-1: A Center Diameter Circle centered on the origin with a diameter of 50 mm

Notice that the line you just drew is blue, which indicates that it isn't completely constrained. To make sure the line passes through the center of the circle, select both the line and the centerpoint of the circle by holding down CTRL on Windows or COMMAND on Mac. Then, from the Sketch Palette on the right-hand side of the Fusion 360 window, scroll down to Constraints and click **Coincident**, as shown in Figure 4-2. A *coincident* constraint forces your selections to align. The coincident should lock the centerpoint onto the path of the line segment, though it can still move anywhere on that path.

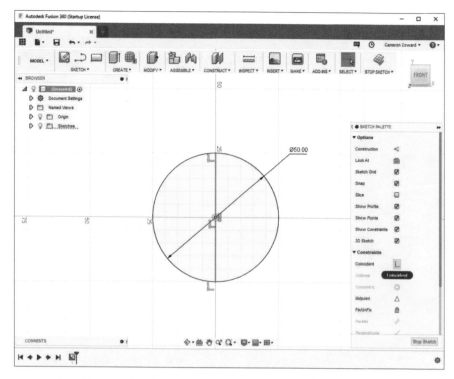

Figure 4-2: Adding a coincident constraint will fully constrain the circle sketch.

The selected profile shouldn't cross the axis of revolution (the y-axis here) but it can touch the axis. Right now, the circle is crossing the axis of revolution, so you need to trim it. You can either select half the circle in the feature options or select the **Trim** tool and click somewhere on the left side of the circle.

The part of the circle you're removing should be highlighted in red. The Trim tool cuts the line off at its nearest intersection points. In this case, the nearest intersection points are where the vertical line meets the circle. You should be left with a half-circle that touches the axis of revolution.

Revolving the Circle

Now you can select the Revolve tool. Choose the Profile and then choose the axis of revolution. In order to make sure you successfully selected the y-axis, click the arrow next to the Origin button at the top left of the screen. The arrow should reveal the default reference geometry (shown in Figure 4-3), which includes the automatically generated origin, axes, and planes. You can select **Y** from there.

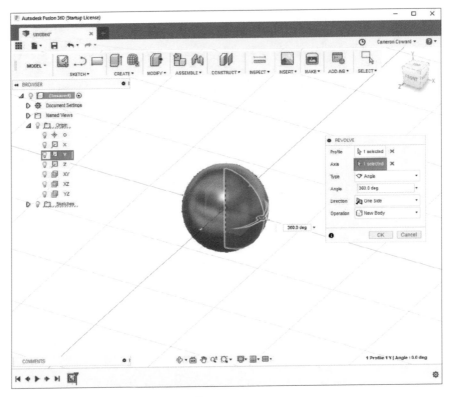

Figure 4-3: Selecting the y-axis manually

The Type setting should be either Angle at 360 degrees or Full; for the Operation setting, select **New Body**. You now have a shiny new sphere like the one shown in Figure 4-4!

Figure 4-4: A basic sphere

Modifying the Sphere

Now that you have a basic sphere, spice it up by adding a couple of features. First, use Extrude to put a hole through the center of the sphere, down its vertical axis.

Because this is a sphere and doesn't have a flat face, instead of sketching on a model face, you need to sketch on the existing Top plane—that is, the x-z plane, which is created by default. Create a new sketch and then choose the **x-z plane** from the Origin folder on the left-hand side of the window (shown in Figure 4-5). Then draw a circle centered on the origin with a diameter of 15 mm and execute a *cut extrude*, choosing a value of **All** for the Extent option. Select **Two Sides** for the Direction so it will cut all the way through the sphere.

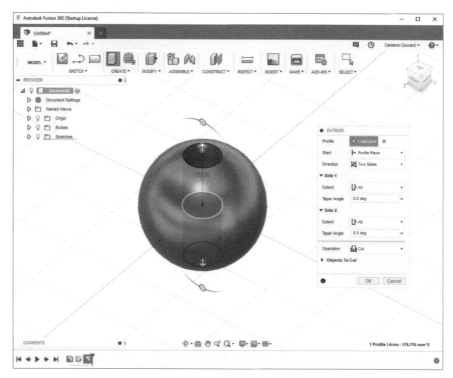

Figure 4-5: Extruding a 15 mm hole through the sphere's vertical axis

You should now have an object that looks like a bead; the edge looks a little rough, though, so go ahead and add a chamfer feature to both openings of the hole. You'll find the Chamfer tool under the Modify menu. It's used to blunt a selected edge.

You define a chamfer by specifying the distance of the cut from the selected edge—either two different distances, two equal distances, or a distance and an angle. In this case, use two equal distances. Enter a distance of 2 mm and finish the feature so that your model looks like Figure 4-6.

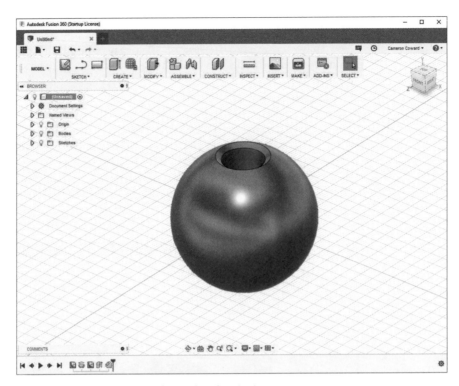

Figure 4-6: The hole now has fancy chamfered edges.

Now you should understand the importance of reference geometry. Next, you'll finally model something useful!

Modeling a Decorative Pencil Holder

The craft of 3D CAD modeling is most exciting when you're designing items you can actually use. Maybe you're planning on 3D printing your models, or CNC milling them, or even sending them out for manufacturing.

In this section, you'll learn to model a basic decorative pencil holder. You'll be using the features you've already learned about, along with a couple of new ones, like Arc and Shell. If you'd like, you could 3D print this model when you're done and put something on your desk that makes your co-workers envious of your new skills.

Creating a Simple Box Feature

Begin by sketching a 75 mm × 75 mm square on the Top plane (x-z plane). Using the Center Rectangle option, center the square on the origin. Then extrude the profile you've created—100 mm up—to create your base feature, as shown in Figure 4-7.

Next, you'll create a Revolve feature.

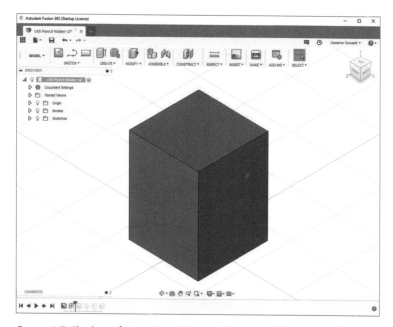

Figure 4-7: The base feature is a 75 mm × 75 mm × 100 mm extrude.

Sketching an Arc

To create a Revolve feature on the Front plane (x-y plane), first draw a sketch that looks like the one shown in Figure 4-8. Constrain the arc so that it's tangent to a line that's at an 80-degree angle coming from the bottom. Be sure to use the existing y-axis as the axis for the Revolve tool.

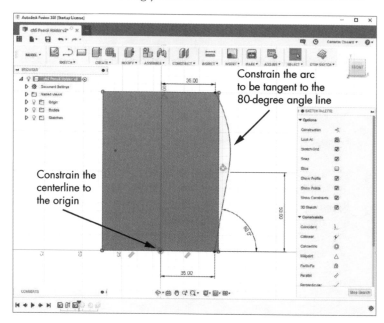

Figure 4-8: Draw the sketch as shown, paying special attention to the constraints.

The only new tool you need to use is the Arc. Draw the arc from the angled line to the top horizontal line; then select the angled line and the arc and give them a tangent constraint from the Sketch Palette. You can use the 3-Point Arc, or you can use the Tangent Arc to save yourself the second step of adding the tangent constraint manually.

Revolving the Arc Feature

With the sketch finished, you can now create the Revolve feature. You do this the same way you've done before—by selecting the sketch you just drew as the profile and making the y-axis the axis of revolution. This time, however, change the Operation type to **Intersect**, as shown in Figure 4-9.

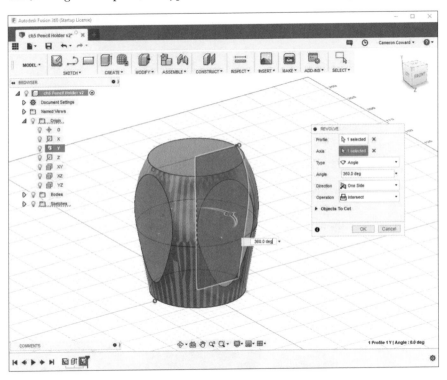

Figure 4-9: Using the Revolve feature with the Intersect option.

The Intersect type leaves behind *only* the geometry where the existing solid and new solid overlap each other. In this case, the solid that the Revolve feature would have created doesn't quite reach the corners of the box that the Extrude feature created, so the Intersect operation removes that part of the model—the corners where there is no overlap. You should now be left with a solid that looks like Figure 4-10.

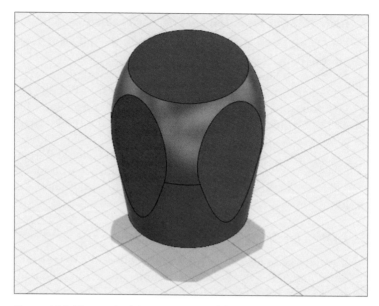

Figure 4-10: The result of the intersection of the Extrude and Revolve features

You now have an interesting shape, but those edges don't really mesh together well—visually, it's just a bit jarring. Chamfers and fillets are useful for smoothing out abrupt edges like that, and they give your model a more refined aesthetic. To improve the appearance, add 5 mm chamfers or fillets to the bottom edge as well as to each of the four teardrop-shaped edges, as shown in Figure 4-11.

Your pencil holder should now have round edges.

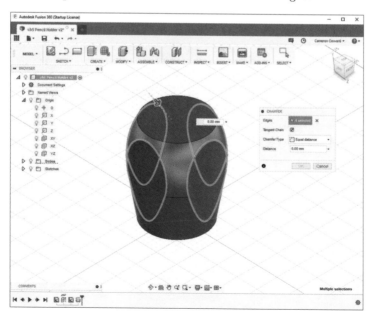

Figure 4-11: Chamfers and fillets are great for improving the finish of a model.

Hollowing Out the Model with the Shell Feature

Finally, add a Shell feature to hollow out the model; that way, you can actually put pencils inside it.

Choose **Shell** from the Modify drop-down, make sure Tangent Chain is unchecked, and select the top face. This tells Fusion 360 that this is the face you want to be open. The Direction setting should be set to Inside, and the Inside Thickness setting, which is the thickness of the walls, should be 5 mm.

Click **OK**, and you're done! Your model should have an open top and a hollow interior, with 5-mm-thick walls all around. Play around with the Extrude, Revolve, and Chamfer features to tweak the design to your liking.

Printing the Model

If you'd like to 3D print your design, choose **3D Print** from the Make drop-down menu. For the Selection option, choose the solid body of the model you want to print by clicking on your model. You can set the quality of the mesh with the Refinement option, which determines how many triangles are used to form the mesh. Usually, the only reason *not* to use the highest settings is to keep the file size small. Uncheck **Send to 3D Print Utility** if you just want to save the STL file to print later. Leave it checked to automatically export the STL file to the slicing software of your choice.

Exercises

You should complete the following projects to practice the skills you've learned so far. The tools and features covered up to this point in the book will be enough for you to do each of the projects.

Remember, there is no right or wrong way to model something—even though there are best practices. The steps you take to create these may not be the same steps someone else takes; what matters is the final result and that you understand what you did and why you did it.

The actual dimensions of these models aren't important. They're just jumping-off points for you to practice and test what you've learned. Feel free to alter the designs or add to them as you see fit!

Money Clip

Try modeling the simple money clip shown in Figure 4-12; then try adjusting the design to personalize it or to make it more functional.

Figure 4-12: A simple clip for holding your money!

Shirt Button

Shirt buttons pop off so easily, and who can ever remember where they put those extra buttons that come with shirts? Now you can 3D print your own replacement buttons! The model in Figure 4-13 has a concave top face for a little extra difficulty.

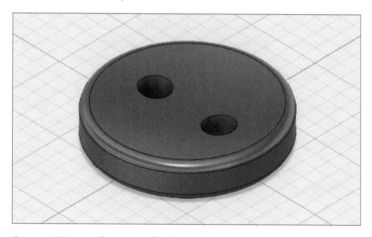

Figure 4-13: A replacement shirt button

Once you've mastered this button, try replicating the buttons on a shirt you already own.

Electronics Leg Bender

Do you ever work on electronics projects and find yourself struggling to bend the legs of components to nice, consistent lengths? The handy tool shown in Figure 4-14 can fix that. It has slots to hold components like resistors or LEDs so you can bend their legs to your desired length. Try customizing it to match the spacing on your perfboard.

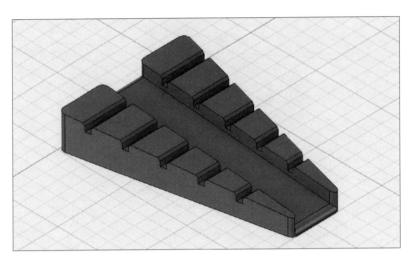

Figure 4-14: Use this tool to bend the legs of electronics components.

Summary

In this chapter, you developed some important new skills and expanded your modeling vocabulary to include more tools. Throughout the rest of this book, you'll learn about increasingly more advanced tools and techniques, but you can already complete a lot of projects using just what you've learned so far.

The vast majority of the models you make will be composed of features like these, which seem simple at first glance but are so versatile that you can use them to create an incredible variety of geometry. Try the following exercises to model some useful parts; then experiment with using your new skills in a modeling project of your own.

5

MODELING ASSEMBLIES

At some point, you'll get bored of modeling designs that consist of a single part. In this chapter, you'll learn how to use Fusion 360 to create more complex and multipart mechanisms, called *assemblies,* which can be made up of two or more *components* (parts) or even multiple subassemblies. A car's engine is an example of an assembly.

Fusion 360 offers a few ways to create assemblies. You can make and assemble multiple solid bodies within a single Fusion 360 file. You can also combine separate files to form an assembly. In this chapter, you'll learn both methods as well as how to combine them.

The method you use depends on what you're trying to accomplish, as well as what makes the most sense to you on an intuitive level. For some people, having each file consist of a single solid body (one component) seems like the most natural way to structure things. Others prefer to have a single file that contains all of the solid bodies—converted into components—for their assemblies. Using a single file may be simpler, but it provides less flexibility. Unless you have a good reason not to, keep components as separate files, because doing so makes the components easier to edit and helps with organization and collaboration.

In this chapter, you'll also learn about reference geometry, an invaluable tool for creating complex models and the relationships within those models, which is necessary for advanced modeling.

Converting Solid Bodies into Components Within a Single File

Components allow you to build assemblies in the same file as you build your bodies. Classically, most CAD software packages force a user to work on parts and assemblies in separate files; you'd choose between a part file or an assembly file. Fusion 360 does not explicitly differentiate between file types, so you can edit both parts and assemblies in the same file. Users create and manipulate bodies within components, which maintain their own coordinate systems and can be created in either the same file or separate files.

Here are some common user interface terms:

Joint   A physical relationship between components.

Component   Holds local coordinate system, bodies, features, and sketches.

Body   Holds geometry with a type of construction.

Feature   An action on a design and the parameters of that action.

Parameter   A named value. Designs are recalculated when a parameter is altered.

I'll begin by showing you how to create and assemble components within one Fusion 360 file. Going this route has two primary benefits. First, since you'll store all the parts in one file, they are easy to keep track of. Second, by splitting a single solid body into two, you can ensure an exact fit between components.

To learn this technique, you'll model a box with a lid, which together form an assembly. Begin by opening a new file and modeling a rectangular prism that is 100 mm wide, 75 mm deep, and 50 mm tall. For a little panache, go ahead and add 5 mm fillets to all of the edges except the bottom four (see Figure 5-1).

To turn this object into a box and a lid, you need to divide it into two pieces. You can do this easily with the Split Body tool. First, though, you need a way to define where that split is going to happen.

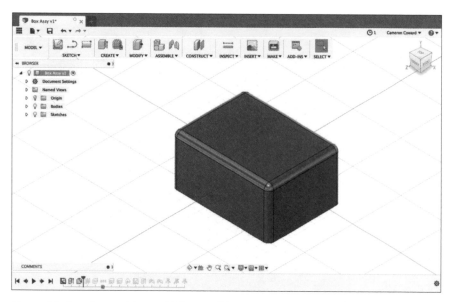

Figure 5-1: Start by modeling a 100 mm × 75 mm × 50 mm rectangular prism.

Splitting an Object Using a Construction Plane

One way to tell the Split Body tool where to cut is to sketch a simple line across the front or side face. This method is particularly useful when the split isn't a straight line. You could, for instance, split a model along an arc.

Because we just want a simple straight cut, we'll use a faster method: cutting along a *construction plane*. Construction planes, axes, and points are all examples of reference geometry. (Reference geometry is used during the modeling process as a guide, but it isn't a physical part of the model.) You might use a construction plane to mirror a feature from one side of the model to the other, or you might use a construction axis as the axis of revolution for a Revolve feature.

In this case, you'll create a construction plane, which tells the Split Body tool where to divide the solid body. Choose **Offset Plane** from the Construct drop-down menu. The Offset Plane tool creates a construction plane parallel to an existing plane of your choosing. In the "plane" box, select the top face of the box. Next, in the "distance" box of the drop-down menu, decide how far the new plane should be from the old plane. A positive value creates the new plane above the original one, whereas a negative value creates it below. For our purposes, enter **–15.00 mm**, as shown in Figure 5-2.

Now you have everything you need to divide the model into two solid bodies. Select the **Split Body** tool from the Modify drop-down menu. When the tool asks you which body to split, you can click anywhere on the model. It will also ask what to use as the splitting tool, or the place where you want to separate the model. In this case, the splitting tool is the construction plane you just created (but it could also be a sketch or other reference). Select the construction plane and then finish the split, as shown in Figure 5-3.

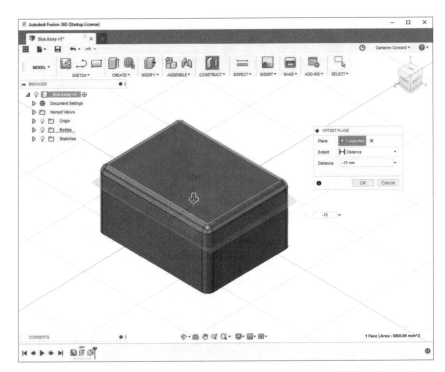

Figure 5-2: Use Offset Plane to create a construction plane 15 mm below the top face.

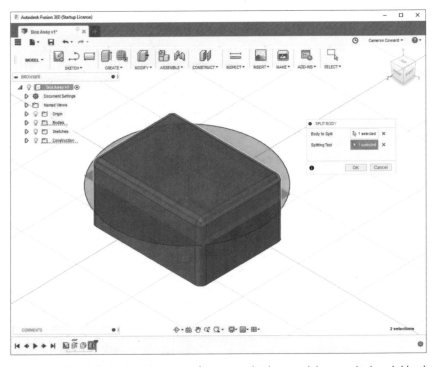

Figure 5-3: The Split Body tool uses a reference to divide a model into multiple solid bodies.

Your model now consists of two separate solid bodies (one for the box and one for the lid). You no longer need the reference plane, so you can hide it by clicking the corresponding light bulb icon in the **Construction** section of the Component Browser (on the left). However, *do not* delete the construction plane because the Split Body feature is based on it. If the plane were deleted, any feature that depends on it would be invalidated, including the split body you just made.

Creating Components

Although the two bodies in our model are split, they're not yet defined as components. To convert all of the bodies into components at once (see Figure 5-4), right-click **Bodies** in the Component Browser and choose **Create Components from Bodies**. You could also do this by right-clicking each body individually.

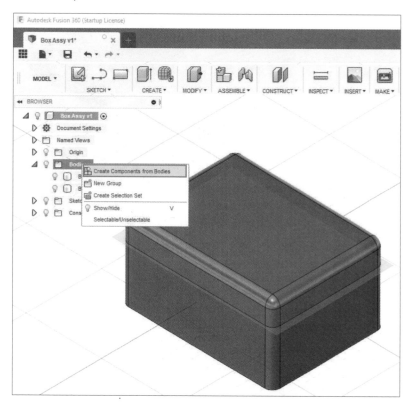

Figure 5-4: All of the bodies can be converted into components in a single step.

The Component Browser should now have two components listed: the lid and the box. If you click the drop-down arrow for each of these, you should see that each component now has its own origin, planes, and bodies. You've effectively put two independent models within the same file. You can move each individually while still retaining the original coordinates of each.

Hollowing Out the Box and Creating a Lip

Right now, each component is solid, but to store anything in the box, you need to make it hollow, as shown in Figure 5-5. You already know how to do this with the Shell tool, so give the box a shell thickness of 3 mm and the lid a thickness of 6 mm. To work on each component individually, hover over it in the Component Browser and click **Activate Component**. To edit both components at once, hover over the top level of the Component Browser and select Activate Component.

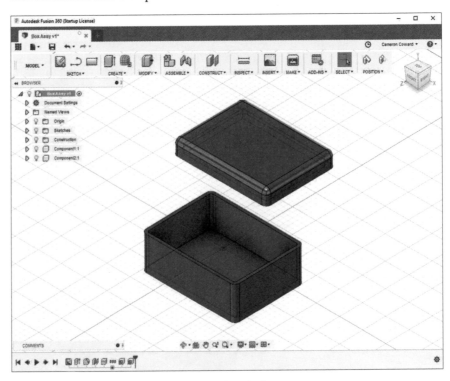

Figure 5-5: Shell each Component to make them hollow.

The lid component should have a thicker shell than the box because you don't want the lid to slide off when you close the box. To keep it from doing that, you'll need to give the lid a lip that fits into the bottom box. Using a thicker shell provides extra material to make that lip. Activate the lid component and rotate it so you can see the inside, then create a sketch on the bottom face (where the lip will be). Next, offset the outside edge by 3.5 mm and extrude the *inside profile* by 2.5 mm, as shown in Figure 5-6.

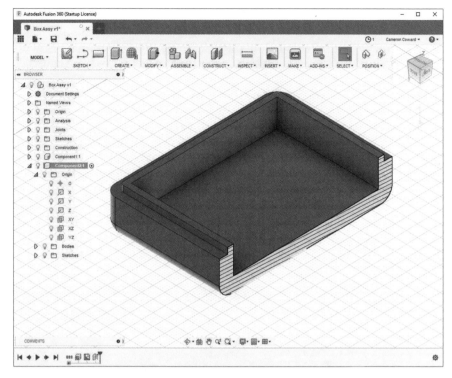

Figure 5-6: A cross-section view showing the extruded lip on the lid component

You should leave a 0.5 mm gap between the edge of the lip and the box, because when you're designing parts that mate together, it's important to think about how they'll actually fit in the real world. If these two parts were modeled to have zero gap, then the actual objects might fit together too tightly (or not at all). In engineering terms, the room you allow for some error is called *tolerance*. This 0.5 mm tolerance helps ensure that the components will fit together, even if they aren't manufactured perfectly.

How much tolerance to give your mating parts is a bit tricky. If you want the parts to fit together snugly, you might not give them any extra tolerance at all. If they need to be able to move freely, you might give them a lot. Determining the exact amount takes experience, trial and error, and knowledge of what you can expect from the fabrication techniques and materials used to make the part. High-quality machining can be very precise, so it's usually safe to use a very tight tolerance in those cases. However, 3D printing is inherently imprecise (particularly hobbyist fused-filament fabrication printing), so loose tolerances are more appropriate there.

Defining Relationships

When building assemblies, you'll usually want to create relationships between the components. For example, when modeling a piston and a cylinder, you need to explicitly state that the piston is centered in the cylinder. You also need to specify its orientation, as well as the distance it can travel in the cylinder.

To define the box's relationship to the lid, you'll first need to *ground* one of the components. This should lock that component in space, allowing the other components to move in relation to it. Generally, you'll ground whichever component you think of as "the base." Right-click the box component and click **Ground**.

Next, you'll use joints to define the motion of the lid in relation to the box. Joints either restrict or allow a certain kinds of movement; that piston in the cylinder from the previous example would use a cylindrical joint to allow movement along the central axis, but not perpendicular to it.

To allow our box to open and close, let's keep the lid centered on the box while still allowing it to move up and down.

From the Assemble drop-down menu, select **Joint** (or use the **J** shortcut). Next, choose a component and a reference point on that component. Usually, this is part of a face or a centerpoint. Select the centerpoint of the side faces of each component (see Figure 5-7). Then, click **Flip**, if necessary, and choose **Planar** as the **Motion Type**.

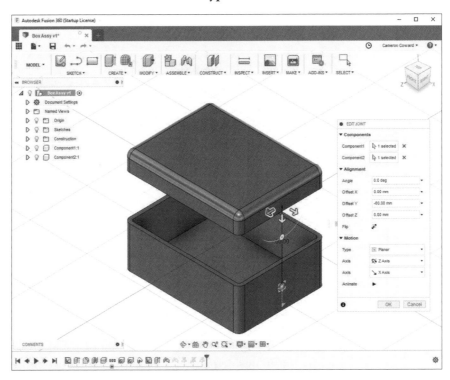

Figure 5-7: Add a planar joint to the side faces of the components.

Repeat this process to add a second planar joint to the front faces. Now you should be able to move the lid freely up and down, but not side to side or front to back.

Your first assembly is finished! If you want, you can export each component as an STL file and 3D print your own box using this model.

Assembling Components from External Files

While you could model and assemble all of your parts in a single file, as you did for the box, it can be useful to separate the components into their own files. Here are a few reasons why:

- When you're working with very large assemblies consisting of many parts, it makes keeping the components organized easier and keeps file sizes manageable.

- It allows you to collaborate with others so that each member of an engineering team could work on a different component of the assembly simultaneously.

- Most assemblies have duplicate parts. If you were designing a car, you wouldn't want to model every M5X100 machine screw individually. Instead, you'd want to model one screw and drop as many instances as you needed into your final assembly file.

To learn how to create assemblies from separate files, we'll create a door hinge with two identical sides. You'll create a single model and import two copies of it into an assembly file to complete the hinge.

Making the Hinge Barrel and Mount

Let's start by making the barrel of the hinge. This should be a hollow cylinder, with the sketch on the top plane, extruded symmetrically. Make the outer diameter (OD) 12 mm, the inner diameter (ID) 8 mm, and the total length 100 mm, as shown in Figure 5-8.

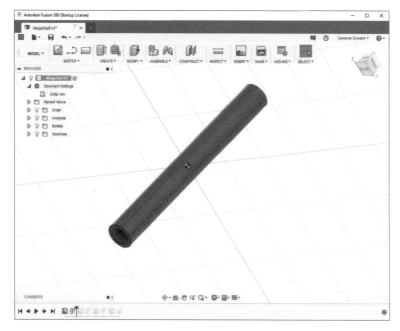

Figure 5-8: Extrude a cylinder with the following dimensions: 12 mm OD, 8 mm ID, 100 mm long.

Next, create a flat face for the hinge mount. Start by sketching on one of the end faces of the cylinder. Move away from the cylinder and complete three sides of a rectangle that is 56 mm long and 3 mm wide.

There is no need to trim the overlapping lines—they won't hurt anything if you leave them like you see in Figure 5-9. If we were actually producing this hinge, we'd want to give it some space for tolerance. But, for the sake of simplicity, you can just align one line with the centerpoint of the barrel.

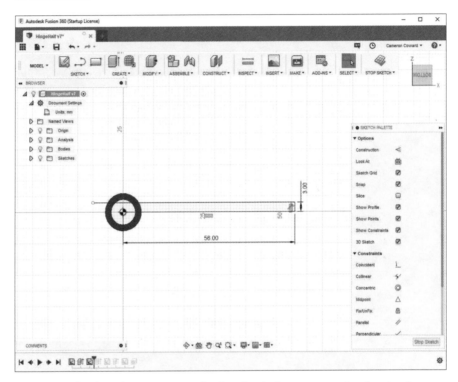

Figure 5-9: Sketch a plane to create a flat face for the hinge mount. Overlapping lines aren't a problem; you can just extrude the closed loop profile.

Once your sketch looks like Figure 5-9, select the rectangular region and extrude it to the other end of the barrel. Your model should now look like Figure 5-10. The next step is to cut out portions of the barrel.

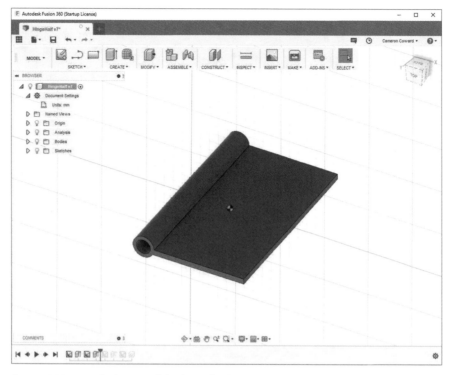

Figure 5-10: Extrude the rest of the hinge body.

Cutting Out the Barrel and Changing Parameters

We'll put two gaps in the cylinder so the halves will fit together like puzzle pieces. Each gap must be 25 mm long (the length of the cylinder divided by 4), 12 mm wide (the OD of the barrel), and 50 mm from the flat edge opposite the barrel. Draw and extrude two rectangles with those dimensions to divide the barrel into four equal sections, as shown in Figure 5-11.

But wait! What happens if the original dimensions change? If, for instance, you alter the diameter of the barrel, the sections you just created will no longer be to scale. This is where you can take advantage of the power of parametric modeling by reusing the dimensions you've already specified.

Go back and edit the sketch you just created. Then, open up the **Change Parameters** dialog from the Modify drop-down menu. Under the Model Parameters tab, shown in Figure 5-12, you'll see entries for each feature you've created. If you expand those entries, you'll be given a list of all the dimensions used to create that feature, as well as the names of those dimensions.

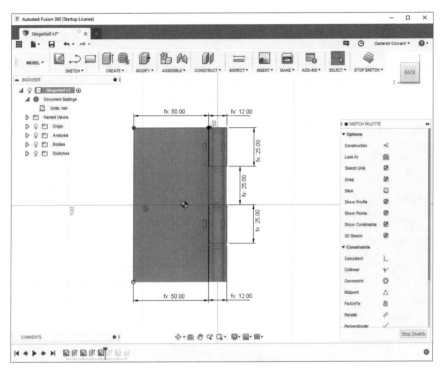

Figure 5-11: Divide the barrel into four 25 mm sections.

Parameter	Name	Unit	Expression	Value	Comme
Favorites					
User Parameters +					
∨ **Model Parameters**					
∨ HingeHalf v7					
∨ Sketch1					
☆ Diame...	d1	mm	12 mm	12.00	
☆ Diame...	d2	mm	8 mm	8.00	
∨ Extrude1					
☆ Along...	d3	mm	100.00 mm	100.00	
☆ TaperA...	d4	deg	0.0 deg	0.0	
∨ Sketch2					
☆ Linear ...	d5	mm	56 mm	56.00	
☆ Linear ...	d6	mm	3 mm	3.00	
> Extrude2					
> Sketch4					
> Extrude3					
> Sketch5					
> Hole1					

Figure 5-12: The Parameters dialog lists all the dimensions you've used in earlier features.

Now you can change certain dimensions so they reference previous features by replacing the numerical entries with values that depend on the dimensions of other objects. For example, because the length of the barrel is named "d3," you can change the expression for the length of the rectangles to "d3 / 4" to make sure they will always measure a quarter of the barrel's length.

At this point, your hinge should look like the one shown in Figure 5-13.

Figure 5-13: The solid should look like this after your cuts are extruded.

Now we just need to create the mounting holes.

Using the Hole Tool

The final step of the modeling process is to give the flat plate some mounting holes, where the screws will go. This step isn't actually important to this tutorial, but it's good to get into the habit of adding details.

You could put these holes in however you like, but I'd recommend a method we haven't used yet: the *Hole* tool. For the hinge to work, the screw heads need to be flush with the surface of the flat plate. This means you'll have to make countersunk holes. Add some sketch points where you think the holes should be on the *inside* face of the hinge. Use constraints or parameter references to place these. This way, their placements will update if the hinge dimensions change.

Finish the sketch and choose the Hole tool from the Create drop-down menu. Select each of the points you just sketched. Change the **Hole Type** setting to **Countersink** and then explore the rest of the settings to get a feel for what they do. Finalize the feature when you think it looks like the mounting holes on a hinge should (see Figure 5-14).

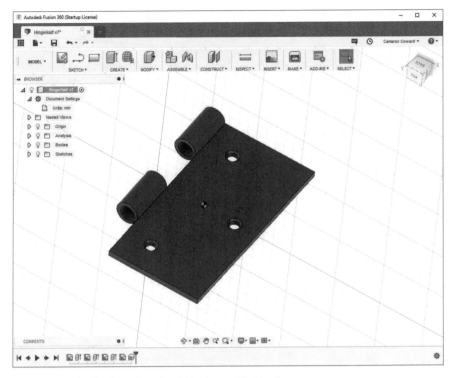

Figure 5-14: The final product should look roughly like this.

Go ahead and save this file as something like *HingeHalf*. Then open a new document and save it as *HingeAssembly*.

When you're in the blank *HingeAssembly* file, open the Project Browser using the Data Panel button at the top left of the Fusion 360 window. Right-click the *HingeHalf* model and choose the **Insert into Current Design** option. Click **OK** in the dialog to place the component anywhere within the *HingeAssembly* document's space. Then repeat this process to insert another copy of the *HingeHalf* model.

Now you can assemble the halves, just like you did in the first part of this chapter. Choose one of the halves to be the ground model and use joints to put the two together, as shown in Figure 5-15. One joint should be cylindrical and the other planar.

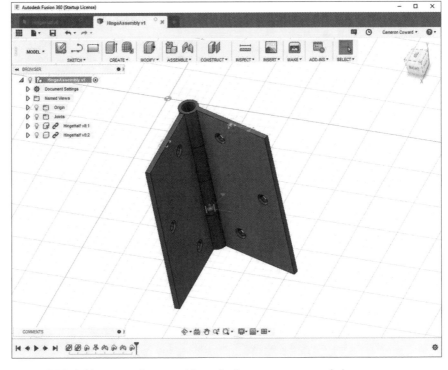

Figure 5-15: Add joints to the assembly so the hinge can open and close.

This should allow your hinge to open and close while staying together.

Combining Assembly Methods

As you can see, both ways of creating an assembly have their benefits. Working within a single file makes it possible to derive several parts from a single base object, whereas creating an assembly from multiple files can save you the trouble of crafting duplicate objects. You can also combine these methods to get the best of both worlds. Let's try this out now by adding a pin to the hinge assembly within the *HingeAssembly* file.

Make sure the top level of *HingeAssembly* is active in the Component Browser and then create a new component by right-clicking *HingeAssembly* (see Figure 5-16).

Model the pin using the tools you've learned so far. Once it's finished, reactivate the top level of *HingeAssembly* to add joints for the pin so that it looks like the model shown in Figure 5-16.

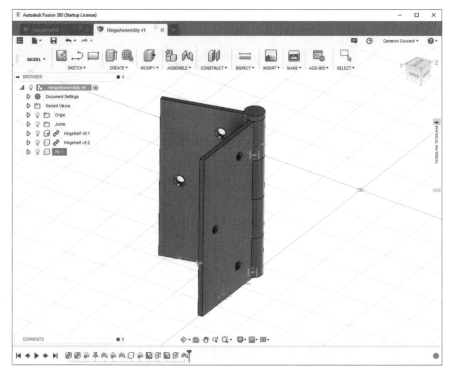

Figure 5-16: Add a pin to the assembly within the HingeAssembly *file.*

Summary

Now that you know how to build assemblies, your projects are no longer constrained to a single part. You can create complex models that contain as many parts as you'd like.

In the coming chapters, instead of telling you which method to use to build assemblies, I'll leave it up to you to choose how best to execute the project.

6

MODELING WITH COMPLEX CURVES

At this point in the book, you've already learned enough to re-create about 75 percent of the mechanical models you can find on a design-sharing site like Thingiverse, even if how to go about it might not always be obvious. Your ability to create complex models depends more on how creatively you can take advantage of basic features than on how many advanced features you know how to use.

That said, you'll still encounter situations where you need to use unusual tools. In this chapter, we'll focus on the Sweep and Loft features. You'll use these to create organic-looking shapes so that you won't be limited to building blocky models.

Sweeps and Lofts

The *Sweep* feature shown in Figure 6-1 creates a solid body by extruding a closed-loop profile along a path. Unlike the path of a regular extrude, the path of a sweep can curve and twist in all three dimensions, which makes this feature great for creating wires, tubes, and handles.

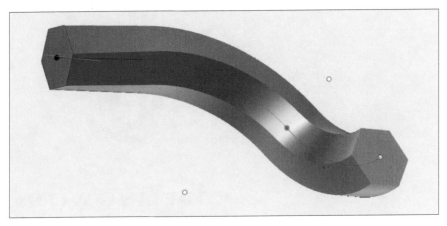

Figure 6-1: The Sweep feature moves a profile along a path to create tube-like bodies.

The *Loft* feature creates a solid body by using two or more closed-loop profiles as cross-sections, which the software then merges into a single solid. Unlike Sweep, the Loft feature doesn't need to use a path; it simply morphs one cross-section into the next, yielding a single seamless body like the one shown in Figure 6-2.

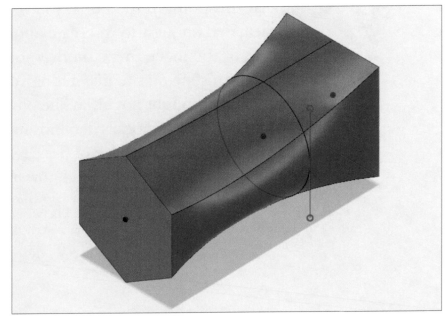

Figure 6-2: This loft stitches together three cross-sections of different shapes and sizes.

Before we use these features, you'll need to familiarize yourself with the concepts of tangent lines and planes, along with perpendicular lines and planes. Both are critical to getting the solid you want from sweeps and lofts.

It's surprisingly difficult to express the concept of tangency. Mathematically, it's a line with a slope equal to the slope of a curve function at a single point, but you most likely understand it intuitively as a line that continues on from a point in a curve. Imagine swinging a marble at the end of string and then releasing the string. The line that the marble flies along would be tangent to the curve of the marble's path at the exact moment you released the string.

This is illustrated by the sketch shown in Figure 6-3. On the left, the line is tangent to the circle—just like the path of the marble after it has been released. On the right, the line is *not* tangent and represents a path the marble therefore *couldn't* follow. The good news is that Fusion 360 understands tangents, and all you have to do is add one as a constraint.

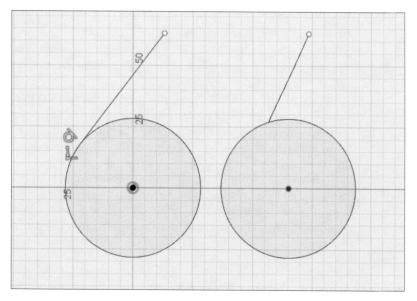

Figure 6-3: Only the line on the left is tangent to the corresponding circle.

Perpendicular lines are much simpler—they're just two lines, a line and plane, or two planes that meet at 90-degree angles.

These concepts are important because both the Sweep and Loft features take relative angles into account. The Sweep feature looks at the angle of the path where it meets the profile, whereas the Loft feature looks at the angles between the cross-sections. In most cases, you'll want to ensure that the paths and profiles are either tangent, perpendicular, or parallel in order to avoid surprises. In addition to the parts of a single sweep or loft, you'll also want to pay attention to how they intersect other features. The model in Figure 6-4 shows a gap created because the ends of the sweep weren't made perpendicular to the block faces.

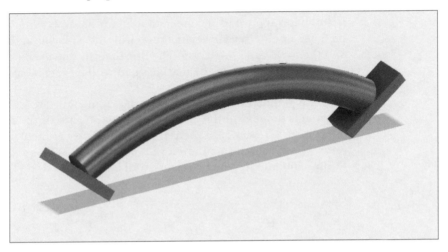

Figure 6-4: The path for this sweep isn't perpendicular to the blocks it joins.

This model could easily be fixed in one of two ways. The sweep path could be made perpendicular to the blocks by either decreasing the arc radius or changing the block angle. Alternatively, if those items couldn't be changed because of some design constraint, the path could be extended further into the blocks to eliminate the gap.

How you choose to handle similar problems will depend on what you're designing, but it's always a good idea to pay attention to the geometric relationships among sketches, paths, profiles, and features. Doing so is not only good design practice but will also save you from modeling headaches in the long run.

Organic Shapes and a Teapot

It's time to learn how to use sweeps and lofts. You'll be modeling a teapot like the one in Figure 6-5. As you can see, it's made of curving organic surfaces you could not have created with extrudes and revolves.

The entire model is made up of only four major features. We'll use a loft for the main body, a sweep for the handle, a second loft for the spout, and a shell to make the whole thing hollow.

Figure 6-5: This teapot is made up of organic shapes modeled with lofts and sweeps.

The Teapot's Body

We'll start by creating the teapot's body using the Loft feature. We'll make three individual sketches and merge them into a single object, as shown in Figure 6-6.

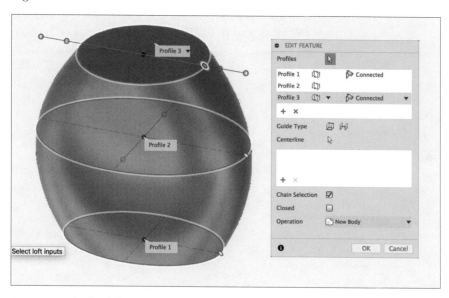

Figure 6-6: The first loft is made up of three sketches.

Feel free to choose your own dimensions, but you should use an ellipse as the base sketch (Profile 1), a larger ellipse as the middle sketch (Profile 2), and a circle as the top sketch (Profile 3), which will make it easy to model a lid later on. All three sketches share the same centerpoint, which should be centered on the origin point.

You can draw the sketch for Profile 1 on the existing x-y plane. For the other two profiles, you'll need to create construction planes. To make those, select **Offset Plane** from the Construct drop-down menu and offset them from Profile 1's sketch plane. When you have your two construction planes, draw your sketches on them.

To complete the loft, match the dialog in Figure 6-6. This loft doesn't require guide rails, so you can leave that section empty. Make sure to check the **Chain Selection** box and set the Operation field to **New Body**. Select each of the profiles you'll use in order and then click **OK**.

The Teapot's Handle

Next, we'll create the sweep that will form the handle of the teapot. You'll create a path for the sweep to follow and then a profile that defines the cross-section of the handle. We'll create the path sketch first. That will give us a reference point to use for our profile sketch, as well as a line we can use to make the profile perpendicular to the path.

Start the sketch of the path on the x-z plane perpendicular to the teapot's base. My path, shown in Figure 6-7, is made up of two arcs, but you can make yours whatever shape you like. That said, you should make sure the path extends into the body of the teapot. If you were to draw it so it stopped at the edge of the body, you'd be left with a gap like the one shown back in Figure 6-4. To make that interaction easier to see, you can switch the Visual Style to Wireframe from the Display Settings menu at the bottom of the screen.

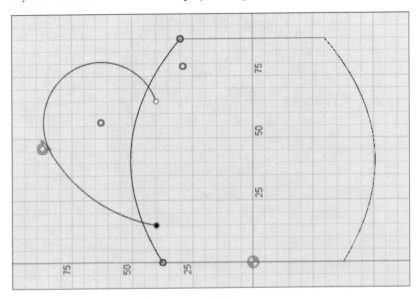

Figure 6-7: Draw a path for the sweep to follow.

You probably noticed you didn't define your dimensions. You should almost always define the dimensions of a sketch explicitly. In this case, though, the actual dimensions aren't very important, because we're just going for visual appeal. You should still constrain the path to make sure it doesn't unexpectedly move. To do that, simply select the lines and then add a *fix constraint* from the same Constraints menu you use to make lines perpendicular or parallel. That should lock the lines in place where they are and change them to green to signify that they're fixed.

Now that you've made your path, you can create the profile for the sweep. To avoid the issue we saw in Figure 6-4, the profile should be perpendicular to the path at the point where they intersect. That's difficult to do, though, considering both arcs end at unusual angles that we didn't specify and wouldn't be able to measure without additional work. So, we'll create a construction plane that's perpendicular to the path.

Fusion 360 actually provides a construction tool specifically for this sort of scenario called *Plane Along Path*, which can be found under the Construct drop-down menu. Choose it and then select the path you just sketched, as shown in Figure 6-8.

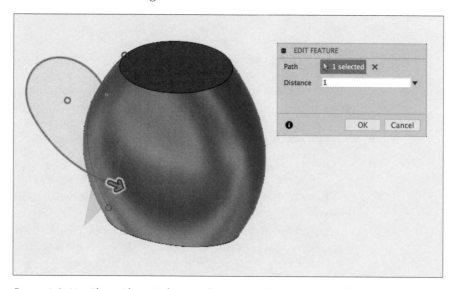

Figure 6-8: Use Plane Along Path to easily create a plane perpendicular to the path at a given point.

Next, set the Distance field, which determines where along the path to put the construction plane. Rather than asking you for a number in millimeters or inches, it asks you for a decimal ratio of the total length of the path. So, "0" would place the plane at one endpoint, "1" would place it at the other endpoint, and "0.5" would place it halfway between the two. In this case, choose either 0 or 1, and the construction plane will automatically appear perpendicular to the tangent at the endpoint of the path.

Now you can be sure that the plane is perpendicular to the path at its endpoint. All you have left to do is draw a profile for the cross-section of the handle on that plane. Unlike the Loft feature, which pieces together multiple profiles, the Sweep feature will simply extend one cross-section along the path. In Figure 6-9, you can see the profile I drew (I hid the body to make it easier to see). Once again, you can make this whatever shape you like.

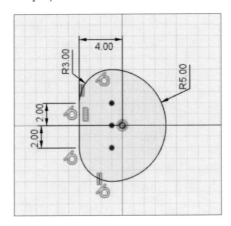

Figure 6-9: The profile sketch for the sweep

When you create the Sweep feature, which is found in the Create drop-down menu, you'll need to select the profile and path you just sketched. The purpose of the other options in the dialog you see in Figure 6-10 isn't quite as obvious. Here's a breakdown of what they do:

Type   This lets you use a guide rail or surface to control the twist of the sweep. It's possible for the sweep to develop an unintended twist, particularly if your path is drawn in three dimensions. A guide can help eliminate this twist.

Distance   Like the Plane Along Path tool, Distance asks you to enter a decimal ratio of the path's length. If you don't want the sweep to go the entire length of the path, you can specify that.

Taper Angle   By default, the profile will remain constant in size throughout the sweep. If you want it to either grow or shrink as it goes, you can specify a Taper Angle setting.

Twist Angle   This spins the profile around the path's axis as it moves.

Orientation   Use this to determine the profile's orientation in relation to the path. Because we drew the profile perfectly perpendicular to the path, we'll want it to remain perpendicular.

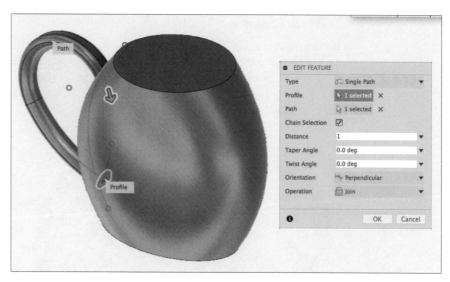

Figure 6-10: The Sweep dialog options are complicated and will usually remain at the defaults.

To give the handle a smooth transition into the body of the teapot, add a couple of large fillets so your model looks similar to Figure 6-11.

Figure 6-11: Fillets are always great for adding smooth transitions.

To do this, just select the edges where the handle meets the body, and Fusion 360 will take care of adding the fillets to the nonuniform edges.

The Teapot's Spout

The next feature, the spout, is the most complex part of this model. It's a loft, but it uses a *guide rail* that acts similarly to the path of a sweep.

As with the sweep path we used for the handle, start by drawing that guide rail. Mine is shown in Figure 6-12. Once you've sketched your own, hold it in place using the fix constraint.

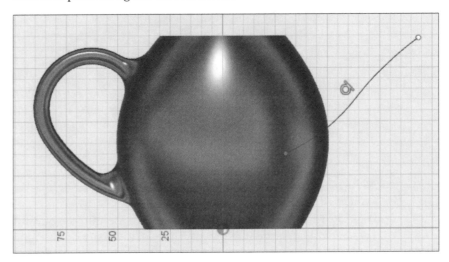

Figure 6-12: The loft's guide rail acts similarly to a sweep's path.

You'll need one profile at each end of the guide rail to form the beginning and end of the spout. Use the Plane Along Path construction plane to sketch those profiles perpendicular to the guide rail. As you can see in Figure 6-13, I made one profile a large ellipse where it meets the teapot body and the second profile a smaller circle at the end of the spout.

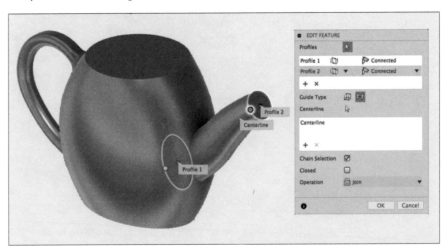

Figure 6-13: Adding profiles on either end of the guide rail to form the spout

To create your loft, start by selecting the two profiles you just sketched. The Guide Type field of the dialog gives two options for the guides: Guide Rails and Centerline. You would use the Guide Rail option if you wanted the guide rail to intersect with the edge of the profiles, and you would use Centerline if you wanted the guide rail somewhere inside the profiles. It doesn't need to be in the exact center of the profiles—that would be very tricky with irregular shapes. It only needs to be located so it's definitely different from a guide on the edge of the profile. Choose Centerline and then select the guide rail you drew.

Hollowing Out the Teapot

To finish up the model, add a fillet at the base edge of the spout and then add a Shell feature for the entire model to hollow out the teapot. When you create the shell, you'll want to select both the top face (where the lid goes) and the face at the tip of the spout.

The thickness of the shell may cause some errors for your model. That's because the handle is fairly narrow, and Fusion 360 gets thrown off when it can't create a shell with a single open cavity. If, for example, the handle is 10 mm wide at its narrowest point, then a 6 mm wall thickness would cause the two walls to intersect, giving you an error. You may need to tweak the thickness and possibly change it to Outside instead of Inside (or even use both). In the end, the model should look something like Figure 6-14.

Figure 6-14: The teapot, in all its organically shaped glory

Exercise

I'll end this chapter by leaving you to model the lid of the teapot on your own using the Sweep and Loft features. It should look something like the one in Figure 6-15.

Figure 6-15: Try modeling this lid using the skills you just learned.

Make sure the bottom of the lid is a circle that will fit into the top of the teapot. The lid should then taper up into an ellipse. Add a small handle on top and use fillets to smooth it all out. You should be able to model this by applying the skills you learned in this chapter.

Summary

The Sweep and Loft features that you learned how to use in this chapter can seem complicated at first, but ultimately they allow you to create organic geometry that is inaccessible with the tools you learned in previous chapters. At this point, you can model practically anything you can imagine. But there are still a few tools that may be useful in specific scenarios. Flip to the next chapter to learn all about coils, threads, and solids that require complex construction geometry.

7

SPRINGS, SCREWS, AND OTHER ADVANCED MODELING

In this chapter, you'll learn some tricks for modeling geometrically complex shapes. You'll learn how to use the *Coil* and *Thread* tools, which are commonly used to make springs and screws. You'll also learn about a way to model complex geometry using *surfaces*, which you'll use to design a 20-sided die.

Modeling a Coil

The Coil tool creates helical, or spiral-shaped, forms. These could also be made with a sweep on a 3D spiral path, but the Coil tool is a lot easier to use. To make a hollow tube for a heat exchange coil, you might use two coils—one for a New Body and one for a Cut.

To get started, select **Coil** from the **Create** drop-down menu. The software then asks you to choose a plane, which you should make perpendicular to the axis of the coil. The centerpoint of the coil's cross-section will start at this plane. Draw a circle, like the one shown in Figure 7-1, to specify the central axis and diameter of the coil.

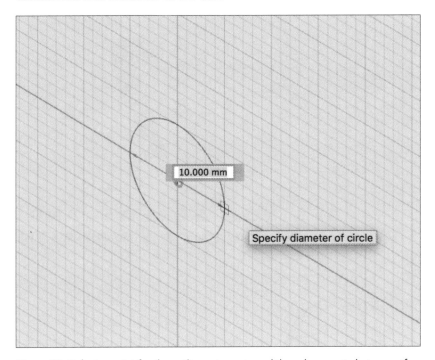

Figure 7-1: Select a point for the coil's center axis and then draw a circle to specify the diameter.

After you've set the circle's diameter, the Coil dialog should pop up, and Fusion 360 should display a model with the default parameters.

Make sure the circle you just drew is selected as the profile. In the **Type** field, choose what kind of dimensions you want to use to create the coil. The default setting asks you to specify the coil's revolution and height. The software will use these dimensions to calculate the coil's *pitch*, or the distance along the axis needed for the coil to make a full revolution. You may want to switch the type if, for instance, you don't care how many revolutions the coil makes but do care about the specific pitch.

Select **Revolution and Height** and then specify how many turns the coil should make in the **Revolutions** field. In the **Height** field, specify the coil's nominal height, which is the distance from the centerpoint of the topmost cross-section to the centerpoint of the bottommost cross-section. Note that it is not asking you for the coil's overall height. If your coil needed to be exactly 100 mm tall, you would subtract the *Section Size* value, which is the diameter of the cross-section, from the Height value. In the **Rotation** field, set the coil to turn either clockwise or counterclockwise.

You already set the diameter when you drew the circle, but you can change it in the Diameter field. If you want the diameter to gradually increase or decrease, you can use the Angle field to give it a conical shape. Use **Section** to set the shape of the coil's cross-section. You can choose to make it a circle, square, in-facing triangle, or out-facing triangle.

By default, that cross-section will center itself on the diameter you select, but if you wanted your coil to fit around an existing part with that diameter, you would set the Section Position field to Outside and choose the part. If you wanted it to fit inside a cylinder with that diameter, you would change the field to Inside.

Take a few minutes to tweak the various parameters and watch how the model changes in response. Figure 7-2 shows the dimensions I chose.

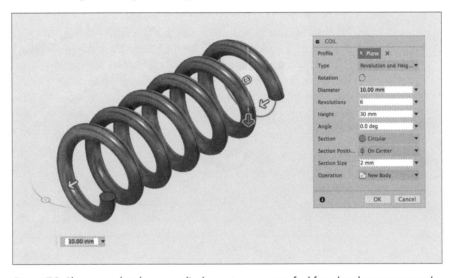

Figure 7-2: Play around with your coil's dimensions to get a feel for what the parameters do.

That's it—you've created a coil! Fusion 360 makes it as easy as that.

Modeling a Screw Using the Thread Tool

Although modeling a screw might seem easy, you'll encounter difficulties if you try to do it with the tools you already know. That's because the spiral-shaped ridge that runs along a screw's body, called the *thread*, has very specific dimensions tailored to the screw's purpose. For example, an orthopedic bone screw will have a completely different kind of thread than a machine screw. When modeling screws, you must define a thread's major diameter, minor diameter, thread pitch, and thread angle, and it's important to get it all right.

On a traditional two-dimensional technical drawing, it's rare for a drafter to bother drawing the threads of a screw. Instead, the drafter will just specify the thread type—for example, 6 g M4×25 mm—so the manufacturer knows what kind of tap or screw to use. If you're planning to buy threaded inserts and screws for your project, you'll probably want to do the same. But if you want to print all your parts with a 3D printer, you'll have to model the threads exactly as you want them to look. In the early days of 3D CAD, that meant precisely sketching the cross-section of each thread and modeling it with a helix feature similar to the Coil tool.

Fortunately for your sanity and patience, Fusion 360 gives you a tool that lets you skip all of that. The Thread tool contains a library of just about every thread commonly in use today. All you have to do is create a cylinder to put the threads on.

To see how it works, you'll model an M4×25 mm screw. This is a very common screw; there are probably even a few of them in whatever computer you're using to run Fusion 360.

Creating the Body of the Screw

Your screw model will have a variable length, so start by creating a Length parameter of 25 mm (**Modify ▸ Change Parameters ▸ New User Parameter**). Then draw a new sketch like the one shown in Figure 7-3 and set the fx: 25.00 dimension to the Length user parameter you made. The "fx:" in the dimension designates that the value has been calculated with a user parameter. Make the screw head 4 mm long and 3.5 mm wide.

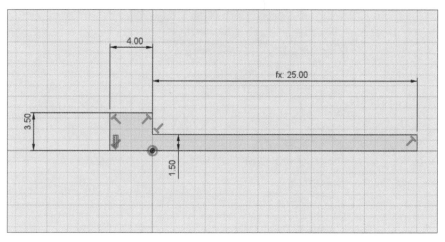

Figure 7-3: Create a sketch like this using your Length user parameter.

I've used the measurements of a standard socket-head cap screw, but the 1.5 mm dimension is completely arbitrary; when you add the threads, the software should automatically resize the width of the cylinder to fit the specifications of the threads. Once your sketch looks like that, revolve it around the bottom line to create a new body.

Like everything else on a standard fastener, this screw has a specific socket size, designed to fit a specific bit. We'll use a 3 mm hex socket, as shown in Figure 7-4.

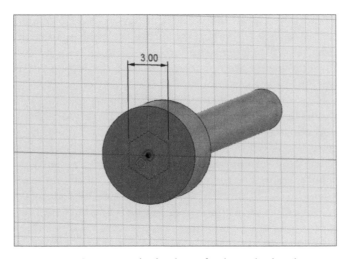

Figure 7-4: Draw an inscribed polygon for the socket head.

To model that, create a new sketch on the top of the head. In the **Sketch** menu, choose the **Polygon** option. This lets you create a polygon with any number of equal sides, like the one in Figure 7-4. In the case of this hex cap screw, you'll need six sides. *Circumscribed* defines it by the diameter of an imaginary circle that touches the midpoints of each edge, while *Inscribed* defines it by the diameter of an imaginary circle that touches the vertices. Choose **Circumscribed** and make the diameter 3 mm.

Modeling the Thread

Now you can create the threads themselves by choosing **Thread** from the **Create** menu. You should see the dialog shown in Figure 7-5.

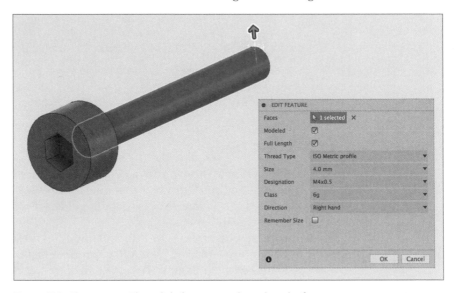

Figure 7-5: Choose your Thread dialog options based on the fastener type you require.

Select the face of the cylinder and then check **Modeled** to make the threads appear there. (If you leave the Modeled option unchecked, Fusion 360 will just store the thread information for a later technical drawing.) Because we want the threads to span the full length of the cylinder, check **Full Length**.

The rest of the options allow you to pick a specific kind of thread. For the **Thread Type**, choose **ISO Metric Profile**. Set **Size** to 4.0 mm and **Designation** to M4×0.5. The Class setting determines how much tolerance to give. We'll use 6 g here. Finally, because this is a standard screw, set **Direction** to **Right Hand**. Almost all fasteners use right-hand threads, which you tighten by turning clockwise, but you might occasionally have a need for left-hand threads, which you tighten by turning counterclockwise.

Your model should now look like a real screw, with built-in threads. As a final step, give the screw's tip a *revolved cut* to make it easier to insert. I gave mine a 45-degree cut that starts 1.5 mm from the center axis, but this is one of the few parts of a fastener without rigid specifications, so you can make the dimensions whatever you'd like. In the end, your model should look something like the one shown in Figure 7-6.

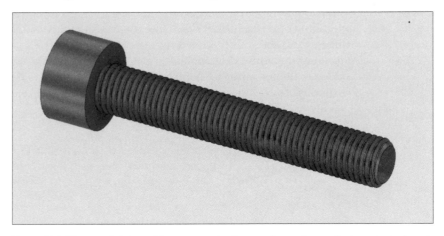

Figure 7-6: Your model should have a revolved cut at the end to make it easier to insert.

Congratulations! You've finished modeling your first screw.

Modeling a 20-Sided Die

Unfortunately, Fusion 360 can't offer a dedicated tool for every complicated design. In this section, you learn how to model a die with 20 sides by using Surfaces, as well as the geometrical concept of golden rectangles, to break the task into manageable steps.

Called regular icosahedrons or, colloquially, d20s, these dice are popular in tabletop role-playing games like *Dungeons & Dragons*. The die consists of 20 identical equilateral triangles joined so that the vertices all touch the same imaginary sphere. Five faces meet at each vertex, as shown in Figure 7-7.

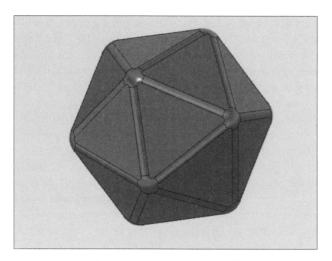

Figure 7-7: You'll be modeling this 20-sided (d20) die.

In geometric terms, the die is quite complex; even using a formula to calculate the length of the triangles' edges would be tricky for people who aren't math whizzes. Fortunately, we can express the geometry in a far simpler way using golden rectangles.

Golden rectangles have dimensions that follow the golden ratio, meaning that their lengths are approximately 1.618 times their widths. Figure 7-8 shows three interlocking golden rectangles placed perpendicular to one another.

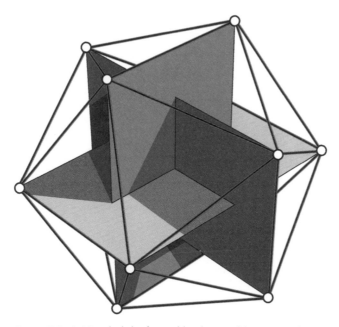

Figure 7-8: A 20-sided die formed by three golden rectangles

When arranged like this, the rectangles' corners form the 12 vertices of the die. The die has 20 sides with 3 corners each, equaling 60. Since 5 sides share the same vertex, there are 12 vertices in total.

When you draw lines between the corners of the rectangle nearest to one another, you form the die's edges. For modeling purposes, we can use those edges to create surfaces for each of the 20 faces. You can then stitch the surfaces together to create a single solid body.

To work with surfaces, switch to the **Patch** workspace, which gives you access to tools in the Create and Modify drop-down menus that you haven't seen yet. Otherwise, this workspace looks the same as the Model workspace.

Use the sketch tools you're already familiar with to draw the three golden rectangles—one on each of the existing planes. Make sure they're oriented the way they appear in Figure 7-9. Give them each a height of 10 mm and a width of 16.18 mm.

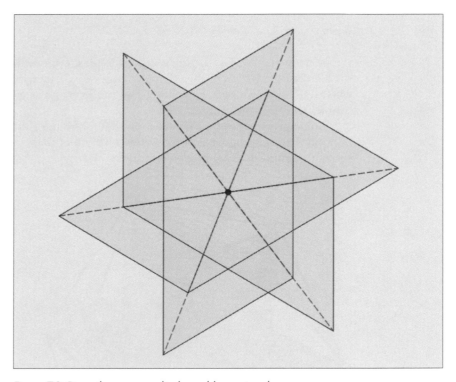

Figure 7-9: Draw three perpendicular golden rectangles.

Next, we need to draw lines connecting the corners of the rectangles to form the edges of the die's faces. To do that, we first need to create construction planes to sketch them on. From the **Construct** drop-down, choose the **Plane Through Three Points** option. Select three points that will make up the vertices of a single triangular face, as shown in Figure 7-10.

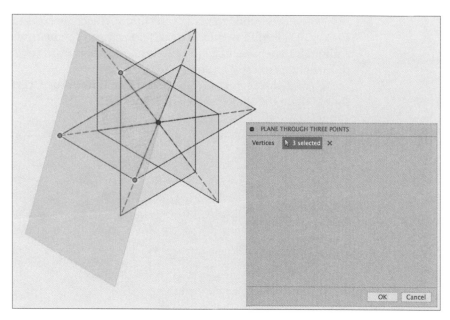

Figure 7-10: Create a construction plane that touches the three vertices of a triangular face.

Now draw a new sketch on the plane you just created. Use lines to connect the same three points you selected for the plane itself. You should end up with an equilateral triangle like the one in Figure 7-11.

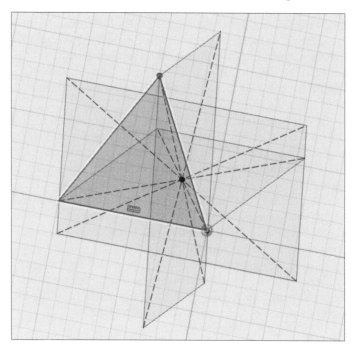

Figure 7-11: Draw a sketch that connects the three points to form a triangle.

Now you're ready to make your first surface. Surfaces have no thickness, so they're not solid bodies like the features you've used so far in this book. Although they have no actual substance on their own, you can connect multiple surfaces to create a solid body.

To model your first surface, select **Patch** from the **Create** menu. This should open the dialog shown in Figure 7-12.

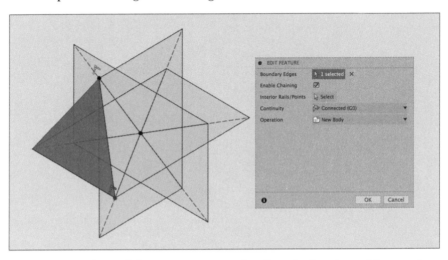

Figure 7-12: Use the Patch feature to create a surface from the three lines of the triangle.

Use the three lines of the triangle you just drew as *boundary edges*. Select **Enable Chaining**. Ignore the Interior Rails/Points option; you won't need these for this project. Set the Continuity drop-down menu to **Connected** and the Operation drop-down menu to **New Body**.

Once you click OK, you should have a single triangular surface—one side of the die. Now, repeat that process 19 more times to generate all of the faces. (Yes, it's a little tedious.)

If you lose track of which points you're supposed to connect, reference Figure 7-8. All edges should be equal in length, so if you end up with an edge of a different length, you've connected the wrong points. In the end, you should have a shape composed of 20 individual faces, as shown in Figure 7-13.

The model now *looks* like a d20, but it's not really a solid body yet. As noted, the surfaces have no thickness. If you tried to export this as an STL file for 3D printing, Fusion 360 wouldn't let you—there just isn't anything there to convert into a mesh.

Fusion 360 provides the *Stitch* feature to turn multiple surfaces into a single solid body. For this feature to work, the surfaces must come together to form an airtight body. If you had only made 19 of the 20 faces, the Stitch feature would fail because of the missing face.

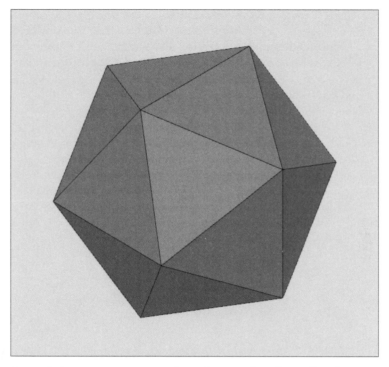

Figure 7-13: Keep creating triangular surfaces until you have all 20 faces.

Since you've got all 20 faces, go ahead and choose **Stitch** from the **Modify** menu; then select all of the surfaces, as shown in Figure 7-14. After you select all 20 surfaces and click OK, Fusion 360 should "stitch" together the surfaces and turn the set into a solid body.

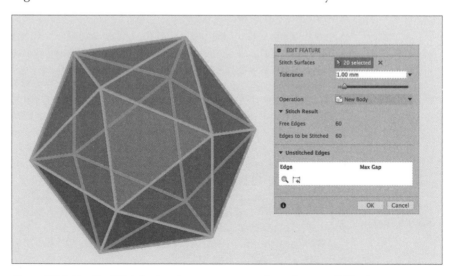

Figure 7-14: Use the Stitch feature to convert your surfaces into a solid body.

The final step is to add fillets to the edges, because nobody wants a d20 with sharp corners. To do that, switch back to the **Model** workspace. Then, use the **Fillet** tool by selecting all the edges and setting a radius of 1 mm. When you're done, your die should look like the one in Figure 7-15.

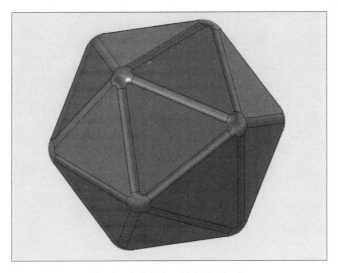

Figure 7-15: Your finished d20 should look like this.

Exercises

Try your hand at the following exercises to get some practice with a couple of minor tools Fusion 360 has to offer.

Changing the Size of Your d20

When you started modeling the d20, I gave you the dimensions of your golden rectangles. I chose those dimensions to make the math easy, but you may want your d20 to be a different size—with, say, a distance of 75 mm between two opposing faces. Of course, you could figure out how the dimensions of the rectangles relate to the distance between the faces of the die and then change the rectangles' dimensions accordingly, but it's far easier to just resize the final model.

If you measure the distance between opposite parallel faces using the Measure tool from the Inspect menu, you'll find that it's something like 15.114974 mm. (This will differ very slightly between faces, because 1:1.618 is actually just an approximation of the golden ratio. Like pi, the golden ratio is an irrational number and continues infinitely past the decimal point. For real-world use, 1:1.618 works fine.)

We want the distance between opposite faces to be 75 mm, so about 4.961967 times what it is now. To make that happen, choose the **Scale** feature from the **Modify** menu and select the model. Keep **Scale Type** set to **Uniform**, as shown in Figure 7-16, and set the **Scale Factor** to **4.961967**.

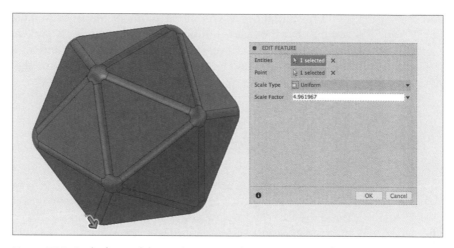

Figure 7-16: Scale the model to make it 75 mm between opposite faces.

Click **OK** to resize the d20. Measure the opposite faces again. The distance should be almost exactly 75 mm (by any reasonable tolerance).

Inscribing Numbers

If you want to actually 3D print your giant d20 to impress the dungeon master at your next D&D session, you have to add numbers, or else you'll find it pretty difficult to use. To add them, choose the **Text** option from the **Sketch** menu and position your text; then extrude it as with any other sketch to cut lightly into the face. I did this to the "20" shown in Figure 7-17.

Figure 7-17: Add numbers to the faces of the die by extruding the text.

Repeat that process for all 20 numbers. Statistically, it doesn't matter how you number the faces—all sides of the die have an equal chance of coming up—but there is a convention for doing so. Pick a random side, and place 1 there. Then, on the opposite side, place 20. On a face adjacent to 20, place 2. Then opposite that, place 19. Adjacent to 19, place 3, and so on. In the end, the sum of any two opposite faces should always equal 21.

Summary

The modeling in this chapter ranged from trivial to pretty darn difficult, but hopefully you've learned that even projects that seem overwhelmingly complex at first glance can be broken down into manageable steps. With the skills you've picked up so far in this book, you should be able to create 3D models of just about anything you can imagine, as long as you take the time to think through how you'll approach it step by step.

Now you know how to use all of the most important modeling tools. In Chapter 8, you'll learn how to create technical drawings, which will be useful if you'd like your models to be built. In Chapter 9, you'll learn how to make nice-looking renders of your models for presentation purposes.

8

DRAFTING

In this chapter, we'll cover drafting, which is the process of creating technical drawings. Ideally, you'd be able to send a file containing your 3D model off to a machine shop or manufacturer to have it made, but in the real world, that's rarely sufficient. Some of the reasons for this are practical. The manufacturer might not use the same CAD software as you do, in which case they couldn't even open your original model file. It's also entirely possible that they won't use CAD at all.

Manual machining is still very common; the manufacturer might not touch a computer when they make your part. Even if they do use CAD, many shops prefer to re-create a 3D model themselves using a system that works better with their machine tools.

But beyond practical concerns, the truth is that a 3D model alone doesn't convey all of the information needed for manufacturing. A model like the one shown in Figure 8-1 only tells the manufacturer what the basic geometry looks like. Other details, like tolerances and surface finish, affect the part too, which is where technical drawings come in. For instance, if you want the shop to clean up the rough edges of a part after it has been machined, you would write "Deburr all edges" on a technical drawing to indicate that.

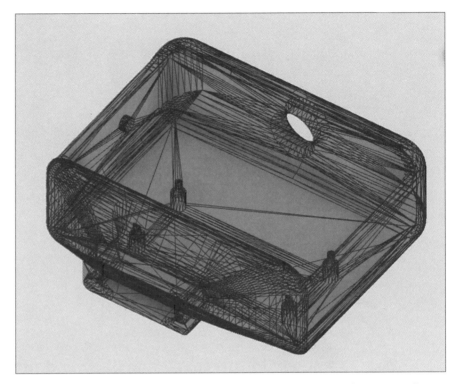

Figure 8-1: STL files like this convey the geometry of the part but not information on how to make it.

Don't despair, though; drafting is pretty easy when you start with a 3D model.

Guidelines for Drafting

Formal drafting has hundreds of rules about everything from the thickness of different lines to the typefaces used for text. Luckily, in the real world, few people care about the vast majority of those rules.

For example, American National Standards Institute (ANSI) rules state that the text height for technical drawings should be 1/8 inch. That's a good rule to follow if you remember it, but if you send a drawing to a manufacturer with a 3/16-inch text height instead, they wouldn't reject the drawing. Drafting as a discipline is full of example s like that. There are rules and guidelines for every single detail of a technical drawing, but at the end of the day, what matters is that the people you send it to can easily and correctly interpret it.

In this chapter, we cover the parts of a technical drawing that impact a manufacturer's ability to make your model. These aspects matter the most, and while there are many of them, Fusion 360 will create most of them for you. You just need to familiarize yourself with them so you'll be able to check the quality of your drawings before you send them out.

Drawing Size

Drawings come in a variety of specific physical sheet sizes, which vary based on the standards the drawing is following. The big architectural blueprints you often see in pop culture, called Architectural E size drawings, are 48 inches by 36 inches, while a standard piece of 8.5-by-11-inch printer paper is called an ASME (American Society of Mechanical Engineers) A size.

That and the ASME B size (11 inches by 17 inches) are the most common, because they can go through regular printers and are easier to store and work with than larger sheets. I recommend sticking with ASME A size sheets, unless you have a good reason to use something else.

Scale

Scale, or the relationship between the size of your drawing and the size of the object, is one of the most crucial details of a technical drawing—and one that many novices overlook. Getting the scale right matters, because someone should be able to measure the drawing if they need to get information about the size of some part. If your part is 1 inch long and has a 1:1 scale, then it should be exactly 1 inch long in the drawing. The drawing would show the part as 2 inches long if you were using a 2:1 scale.

While that's an easy concept to grasp, a few factors could end up making the scale of your drawing inaccurate. The most obvious of these is forgetting to note when different parts of the drawing have different scales. The drawing's *title block*, in the bottom-right corner of the page, shows the drawing's overall scale, which applies to every view unless otherwise stated. Sometimes, though, the drawing will show the same object from multiple views, and it's commonplace for one view to have its own larger scale to more clearly show the detail of some feature.

If an individual view scale differs from the scale of the overall drawing, you must note it in that view's title. Fusion 360 will keep track of those scales for you in the views' Properties dialogs, but you'll have to type them into the text description of the views. The drawing in Figure 8-2 has an overall scale of 1:1, but one of its views has an individually set scale of 2:1.

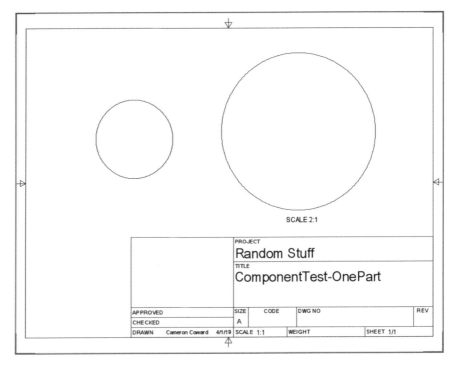

Figure 8-2: The view on the right has a scale that differs from that of the rest of the drawing.

Another common scaling mistake happens while printing a drawing or saving it as a PDF. On most computers, it's pretty common for the print utility to automatically enlarge the document so it fills the entire page or reduce its size so it fits onto one sheet. That will throw off your scale, so make sure the software doesn't do any scaling. You can do that by saving the PDF and printing it at "Actual Size."

Projection Angle

Projections are views drawn using another view for reference. These date back to the days of traditional pen and paper drafting, when drafters would use straightedges to draw guidelines from one view to the next. They also give someone reading the drawing the ability to easily line up features between views. The projections remain locked into place with respect to the front view (the first view you draw) and must maintain the same scale.

The *projection angle* determines how to lay out your views in relation to the front view. You can choose between two options: First Angle Projection and Third Angle Projection. In a First Angle Projection (see Figure 8-3), the view to the right of the front view will show the part as if you were looking at it *from the left side.*

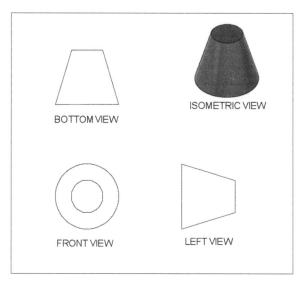

Figure 8-3: A First Angle Projection drawing

A Third Angle Projection (see Figure 8-4) will show the part as if you were looking at it *from the right side.*

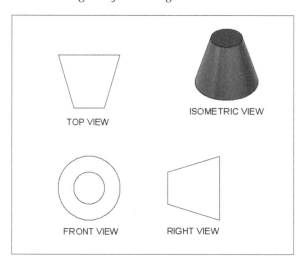

Figure 8-4: A Third Angle Projection drawing

The projection type you're likely to come across mostly depends on where you are. In the United States and Australia, Third Angle Projections are the most common. The rest of the world generally uses First Angle Projections. I use Third Angle Projections in this chapter because I'm American and so was trained on them. The User Preferences window (shown in Figure 8-5) lets you change your projection angle or choose a default.

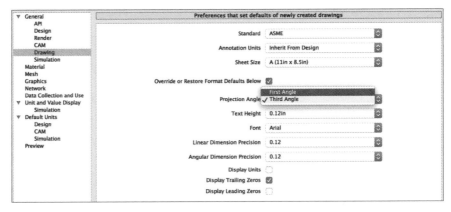

Figure 8-5: Switch your projection style in the User Preferences window.

ISO standards specify First Angle Projections, and ASME standards specify Third Angle Projections.

Tolerances

A tolerance tells the manufacturer how precisely it needs to adhere to the dimensions shown on your drawing (called nominal dimensions). You might have modeled a cube so each side is 50 mm long, but maybe the object doesn't need to meet those measurements exactly. Even the world's most high-tech manufacturing is imperfect; that's just the nature of physical objects. Tolerance lets the manufacturer know how closely it needs to stick to your nominal dimension in order for you to accept the part.

Let's say that 50 mm cube is a stand-alone toy. In that case, it's not important for each side to measure exactly 50 mm, so you might specify a tolerance of ±0.1 mm. That "plus or minus" symbol tells the manufacturer that you'll accept the part if the sides measure anything from 49.9 mm to 50.1 mm. That loose tolerance gives the manufacturer more flexibility in its machining options and makes it easier for parts to pass quality control, potentially lowering what you pay per part.

On the other hand, imagine that your cube has a 25 mm hole going through it where it will mate with the cylinder of a second part. In order to ensure the cylinder fits inside without leaving too much extra room, you might give the hole a much smaller tolerance of +0.01 mm, as shown in Figure 8-6. The machinist will then know that they can make that hole measure anything between 25 mm and 25.01 mm. They'll know not to go below 25 mm, or the cylinder won't fit.

Because tolerances can vary from feature to feature, you can specify them in two places: in the drawing notes or on the dimension itself. The tolerance in the notes is generally the loosest you'll accept for the part as a whole. When a particular feature requires more precision (like the 25 mm hole), you can specify that in the individual dimensions.

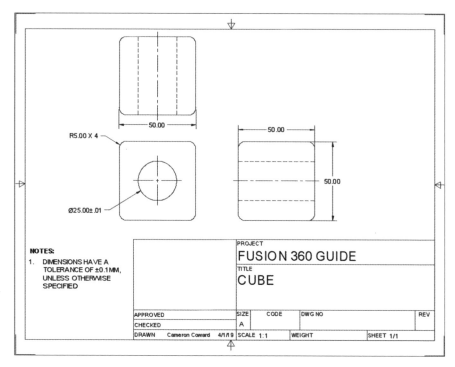

Figure 8-6: The hole in the center of the cube has a tighter tolerance than the other parts.

Line Types

Finally, you'll need to pay attention to the types of lines used throughout your drawings. Once again, Fusion 360 will create these for you, but you do need to know about a few of them, shown in Figure 8-7.

Solid lines represent the edges you can actually see in that particular view. *Regular-interval dashed lines* show the edges of features that are hidden in that view, like those on the other side of the part. You don't always need to show the hidden lines, particularly if the feature is clearly visible in another view. Too much clutter will make your drawing hard to read, so it's a good idea to show hidden lines only if they're actually helpful.

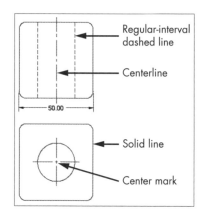

Figure 8-7: Line types in Fusion 360

Certain special lines help illustrate the geometry of a part. The two most common are the *centerlines* and *center marks* on circular parts. You would use a centerline to show the axis of a cylinder and a center mark to show the centerpoint of a hole. Centerlines are usually staggered dashed lines, while center marks are crosshairs that extend just beyond the edge of the hole.

Drafting a Single-Part Drawing

There are two kinds of technical drawings: *part drawings*, which show the specific dimensions of an individual part, and *assembly drawings*, which show how multiple parts fit together. Although it's possible to show part dimensions on an assembly drawing, you should avoid doing so, because the two kinds of drawings serve different purposes. Part drawings tell the manufacturer how to make each of your parts, while assembly drawings tell it how to put those parts together.

Every single one of your parts should typically have its own part drawing. The only exceptions are purchased parts, like bolts, washers, and electronics that you're not making yourself. On the other hand, you only need to create an assembly drawing if you're paying the manufacturer to put your creation together or filing a patent for your invention, in which case the patent office needs to see how it works.

We'll make a part drawing for the hinge you created in Chapter 5. From the Project Browser, open the model part file (not the assembly file). Before you start the drawing, assign a physical material to your model using the **Physical Material** menu option under the **Modify** drop-down. Scroll down to the **Metals** section. Drag and drop **Brass** onto your model, as in Figure 8-8. That tells Fusion 360 the part is made of brass. It will use this information to calculate the part's weight.

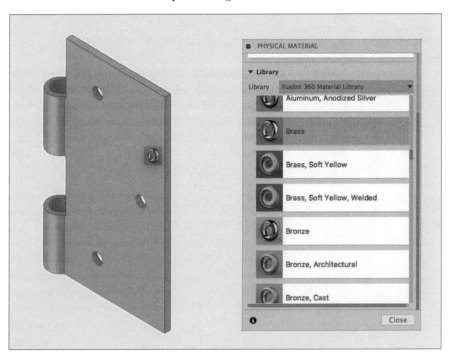

Figure 8-8: Choose Brass from the Physical Material menu and assign it to your part.

Now save the file and switch the Workspace to **Drawing** with the **From Design** option. A dialog should pop up. This gives you some basic options for the technical drawing. Choose the ones shown in Figure 8-9.

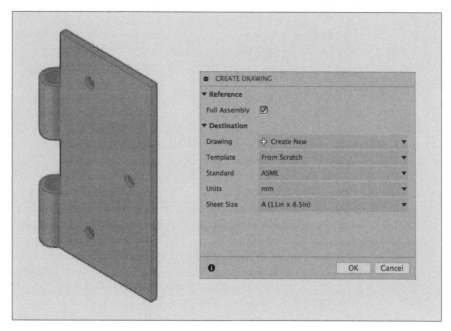

Figure 8-9: Your drawing options should look like this.

Leave **Full Assembly** checked; your file only has one part anyway. In the Drawing field, select **Create New**. You shouldn't have any templates, so leave that field at **From Scratch**. We'll make a Third Angle Projection drawing, so set Standard to **ASME**. In the Units field, select **mm** (millimeters) and set Sheet Size to **A (11in × 8.5in)**.

Creating Your Views

Once you click OK, you should be taken to the Drawing workspace, and Fusion 360 should ask you to place the base view, which it will use to project the other views. The base view can show the part from any side you choose. However, for almost all drawings, you'll want to use a front view for that and then project either a left or right view and either a top or bottom view. You want to choose views that clearly show all of the part's features. Select **Back** from the **Orientation** drop-down menu, because that's the side of the hinge half with the mounting hole chamfers. Otherwise, leave the default settings (shown in Figure 8-10) as they are.

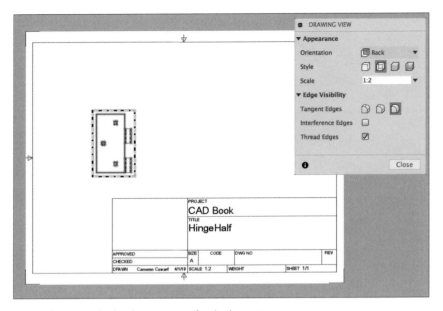

Figure 8-10: Use the back orientation for the base view.

With the base view placed, you can *project* a top view and a right view by using the Projected View tool from the toolbar at the top. Simply click the base view, click again to the right of the base view, and then click a third time at the top of the base view. Press ENTER. It's best practice to place the projected views at approximately the same distance from the base view, as in Figure 8-11, so drag them around if necessary.

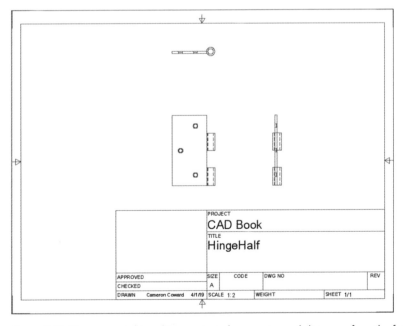

Figure 8-11: Place your right and top projected views at equal distances from the front view.

In Figures 8-3 and 8-4, I labeled the views to show you the difference between First and Third Angle Projections, but you don't have to do that for standard drawings. If you've placed the views correctly, people in the industry will understand that these are front, right, and top views.

For simple parts like this, it's also not necessary to use an isometric view, which shows the part from a three-quarter angle. Generally, you only need that view for parts where the geometry might not be clear from the standard three views. However, if you did want an isometric view, you'd simply add another properly oriented base view. You'd also need to label it and take care not to line it up with another view; otherwise, manufacturers might mistake it for a projected view.

Adding Center Marks and Centerlines

Once you've placed the views, you can begin annotating the drawing. First, place the center marks and centerlines. You can find these in the **Geometry** section of the main toolbar. Add center marks to any holes that appear head on and add centerlines to any hidden holes. When you're done, your drawing should look like Figure 8-12.

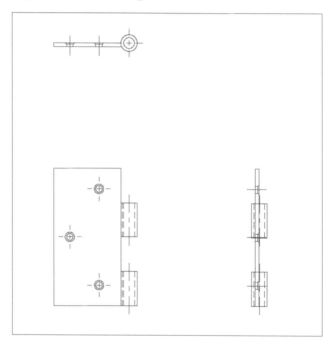

Figure 8-12: Add center marks and centerlines to the holes in your drawing.

Adding Dimensions

Next, add dimensions to your drawing. When dimensioning a drawing, follow these guidelines: make sure you add enough dimensions to accurately define every feature, but avoid superfluous dimensions. If you're unsure, always err on the side of having too many dimensions, but try to avoid cluttering your drawing with unnecessary information.

Standard conventions can help keep things concise. If you've duplicated multiple features (such as holes) in your model, you don't need to specify the complete dimensions on each of them; just write "X3" next to the diameter of one hole. To indicate the diameter and angle of a hole chamfer, put the countersink symbol (∨) before your measurements. When two features are clearly aligned, you only need to add dimensions for one of them. Using Figure 8-13 for reference, add the dimensions to your drawing. You can double-click a dimension to edit it after it has been placed.

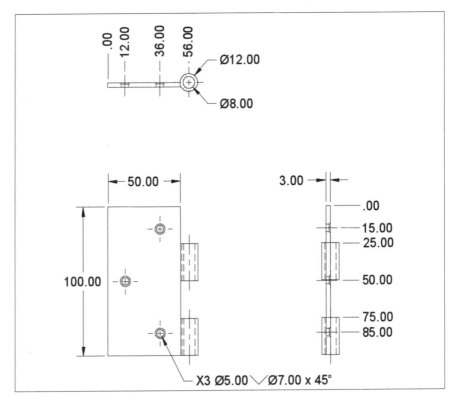

Figure 8-13: The drawing with all the necessary dimensions added

You may struggle with adding some of the dimensions, particularly the "X3 Ø5.00∨Ø7.00 x 45°" hole dimension and the two columns of dimensions that start with ".00". The hole dimension requires that you edit the text itself. First, add the "X3" before the Ø5.00 that Fusion 360 gives you. Then, add the countersink dimensions by using the *Insert Symbol* tool from the dialog.

The lined-up dimensions in the top and right views are called *ordinate dimensions*. The ".00" signifies the starting point, and each of the other values indicates the feature's distance from that point. To add ordinate dimensions, just choose the tool from the **Dimensions** part of the toolbar, select the starting point, then click each of the points you want to create dimensions for.

Ordinate dimensions make a drawing cleaner, less cluttered, and easier to read. They are also important for avoiding a problem called *tolerance stacking*.

As I mentioned earlier in this chapter, you should specify some amount of tolerance for every dimension to give the manufacturer wiggle room when it's fabricating the part. But when the dimensions for one feature are based on those of another feature, the tolerances can add up. Look at the dimensions of the holes in Figure 8-14. Now, imagine those dimensions all have a tolerance of 0.1 mm.

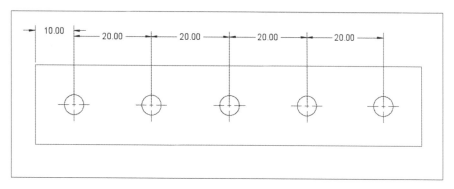

Figure 8-14: Tolerance stacking would make the rightmost hole have a huge variation.

If every distance were at its maximum tolerance, the hole on the far right could be as far as 90.5 mm from the left edge of the part. If every distance was at the minimum tolerance, that hole could be just 89.5 mm from the edge. That gives you a total variation of up to 1 mm, which might make your part unusable. Ordinate dimensions avoid tolerance stacking by using a single hard edge as a reference for each feature.

Adding Text

Now, with all your dimensions in place, you can round out the annotations by adding some notes, as in Figure 8-15. The first note indicates what kind of tolerance you'll accept. The second note tells the manufacturer to deburr the edges after fabrication. To add notes, select the **Text** tool in the main toolbar.

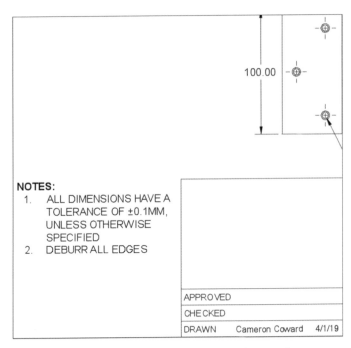

NOTES:
1. ALL DIMENSIONS HAVE A TOLERANCE OF ±0.1MM, UNLESS OTHERWISE SPECIFIED
2. DEBURR ALL EDGES

APPROVED	
CHECKED	
DRAWN	Cameron Coward 4/1/19

Figure 8-15: Add notes to your drawing to give the manufacturer important information.

The final step is to fill out the title block (shown in Figure 8-16), which contains details about the part and the drawing itself. Fusion 360 will automatically fill in some of the information, but you can edit it by double-clicking the title block. If you wanted to, you could place an image file for a logo in the top-left box. The Approved and Checked fields are for the internal document control processes many corporations follow; an engineering manager, for example, might use those to approve a drawing.

GENERIC BRAND, INC.

PROJECT
DOOR THAT OPENS
TITLE
HINGE

APPROVED N/A	SIZE	CODE	DWG NO		REV
CHECKED N/A	A		10021		D
DRAWN Cameron Coward 4/1/19	SCALE 1:2		WEIGHT 152.627 g	SHEET 1/1	

Figure 8-16: A typical title block

In the Drawn field, put the name of the person who created the drawing (you!) and the date it was drawn. In the Project field, you'd generally identify the assembly or subassembly the part belongs to, but there is no hard-and-fast rule about this; call the project whatever you want. In the Title field, write the name of your part. In the Size field, write the kind of sheet you are using (A, in this case). In the Code field, you can specify any regulatory codes that the part and drawing might adhere to. Unless you have reason to stick to a specific code, you can just leave that blank. The manufacturers will use the number in the Dwg No field to refer to the drawing internally and in their communications with you. You can use any number you like, but every drawing must have a unique number, and it's a good idea to come up with some numbering conventions.

In the Rev field, which stands for *revision*, put a letter indicating which version of the drawing you're using. Start with A, and any time you make a change to the drawing, advance to the next letter. If you get all the way to Z, start over with AA, then AB, and so on. Noting the revision is essential for ensuring that the manufacturer is working from the same drawing as you are. If you update drawing number 10021 to revision D, for example, your communications to the manufacturer might say something like "please reference drawing 10021-D."

The Scale field shows the drawing's overall scale. Fusion 360 will automatically set this to whatever you entered as the scale in your initial base view. If that changes, you should modify it in the title block to match. In the Weight field, put the weight of your part. To get that information, go back to your model. Right-click the top level of the Component Browser and check the properties. Finally, complex part drawings can have multiple sheets, so use the Sheet field to indicate which sheet this is.

Once you've filled out your title block, you're done! You can then choose an option from the Output menu (usually PDF) to save the drawing. Remember, if you're going to print your drawing, you *must* make sure to turn off scaling so it prints at its actual size.

Exercise

In this chapter, I walked you through how to create a drawing of a single part, but you'll sometimes have to create assembly drawings, which illustrate how multiple parts fit together. Create one now from the complete Hinge Assembly. It should end up looking something like Figure 8-17.

You'll have to create an *exploded view* of the assembly, which shows how the separated parts fit together. To do this, go to the **Animation** workspace, click **Auto Explode** from the main toolbar, and then save the assembly. If needed, you can reorient the model and save it as a new home view by right-clicking the View Cube. Then, when you create the drawing, choose the **From Animation** option to use that exploded view.

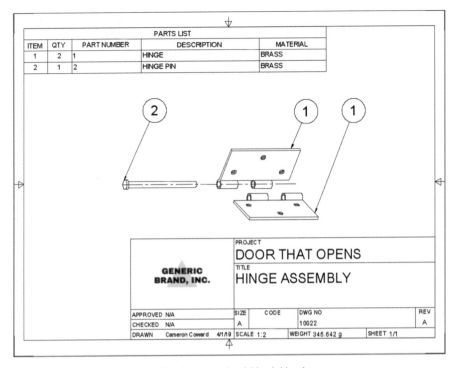

PARTS LIST				
ITEM	QTY	PART NUMBER	DESCRIPTION	MATERIAL
1	2	1	HINGE	BRASS
2	1	2	HINGE PIN	BRASS

GENERIC
BRAND, INC.

PROJECT
DOOR THAT OPENS

TITLE
HINGE ASSEMBLY

APPROVED N/A		SIZE	CODE	DWG NO		REV
CHECKED N/A		A		10022		A
DRAWN	Cameron Coward 4/1/19	SCALE 1:2		WEIGHT 345.642 g	SHEET 1/1	

Figure 8-17: Your Hinge Assembly drawing should look like this.

You also have to add the *bill of materials* (BOM), which is a table at the top of the drawing that shows your parts list. To create that, choose **Table** from the main **Drawing** toolbar. Once you place the table, your parts will automatically populate it. Part Number, Description, and Material are all controlled by the properties accessible from the Component Browser in the Model workspace. Then just place balloons (from the same menu) to label each part in the drawing.

Summary

Technical drawings are very complex, and drafting as a profession takes years to master. Luckily, 3D CAD software like Fusion 360 has made the process a lot easier. Using the skills you've learned in this chapter, you should be able to create clear and professional technical drawings that real manufacturers can use to make your designs. In Chapter 9, you'll learn how to make high-quality renders of your designs for presentations.

9

RENDERING

Rendering is the process by which your computer converts a 3D model into a 2D image that displays on a screen. Technically, Fusion 360 is constantly rendering your 3D model as you're working on it. The same is true in other types of software, like video games. As you play a video game, the game console or computer takes the on-screen 3D assets and renders them for your screen more than 30 times per second.

But when people in the CAD world talk about a "render," they're usually referring to a single photorealistic image or animated video created specifically for presentation purposes, which takes a significant amount of computing power to produce. Fusion 360 is indeed rendering your model many times per second while you're working on it, but it's not creating a high-quality image. Computers have only so much processing power, and in order for Fusion 360 to display your model at a usable frame rate, it has to limit how much time it spends on each frame. That means it creates lower-quality renders—good enough for the design phase but not for presentation.

The more time you give Fusion 360 to work on a single frame, the better the image it creates. Given enough time, it's capable of producing images approaching photorealism. How long that process takes depends on your computer, the quality you want out of the image, and that image's resolution. A quick-and-rough render could take just a few seconds, while a complex, top-quality render could take more than an hour.

Even so, Fusion 360's rendering capabilities are limited. It can certainly produce very nice renders that are more than sufficient for something like a Kickstarter campaign, but it lacks many of the tools found in software designed specifically with rendering in mind. If you need truly photorealistic renders of your model in complex scenes, you can always export your model as an STL file and open it in more advanced rendering software. Figure 9-1 shows a photorealistic scene created using 3D modeling and professional rendering software intended for creating visuals (as opposed to mechanical designs).

Figure 9-1: A photorealistic render created with professional rendering software

This book is focused on Fusion 360, so we'll take advantage of the program's built-in rendering capabilities to create a nice image of the screw we modeled in Chapter 7.

Rendering Your Screw

Go ahead and open the screw model. Right now, it should be a simple flat gray color, which is the default appearance.

There are two ways to change the look of the model. The first is by setting the **Physical Material** option, as you did with the brass hinge in Chapter 8. The second is to explicitly assign an **Appearance** setting, which overrides the appearance of the physical material. The appearance created by setting a physical material will apply to the entire model, but you can assign appearances to either the whole model or individual faces. You might want to give a solid model a two-tone paint job, for example.

In this case, the model is a regular old screw made from a single raw material, so set its **Physical Material** option to **Stainless Steel, Polished**. The model should look a bit nicer now; the color and reflections should look like polished stainless steel. But this is still a low-quality render, because Fusion 360 needs to keep the frame rate high so you can continue to work with the model. If the software were trying to produce high-quality renders continuously, it could take several seconds—or even minutes—to refresh every time you moved the model, rotated it, or modified it.

To dramatically improve how the model looks, switch over to the Render workspace. Immediately, you'll notice that the model looks more realistic than it did in the Model workspace. Before we create our render, we need to make sure the screw is resting realistically on the ground and then prepare an environment to perform the render.

Making the Screw Rest on the Ground Plane

You may have noticed that the shadow on the ground plane doesn't look right at all. The screw is just floating up in the air, which real screws don't usually do. To place the screw on the ground, open the **Scene Settings** dialog from the main toolbar. Click the **Position** button to see the options shown in Figure 9-2. Change the Distance setting, which controls how far the ground plane is from the model, until the ground plane is just barely touching the screw.

Even though the screw head now touches the ground plane, the screw still doesn't look right, because the threaded end is in the air, making the screw appear to balance on its head. That's because Fusion 360 orients the ground plane so that it's parallel with the bottom of the View Cube.

To fix that, we'll need to reorient the screw in relation to the View Cube. Click **Front** on the View Cube at the top right of the screen in order to face the screw head on. Then use the **Free Orbit** option from the Display menu at the bottom of the viewport to rotate the threaded end of the screw down. At this point, the ground plane will rotate along with the screw, but when you're done rotating the screw, you can fix this by right-clicking the **View Cube**, hovering over **Set Current View As**, and selecting **Front** (as shown in Figure 9-3).

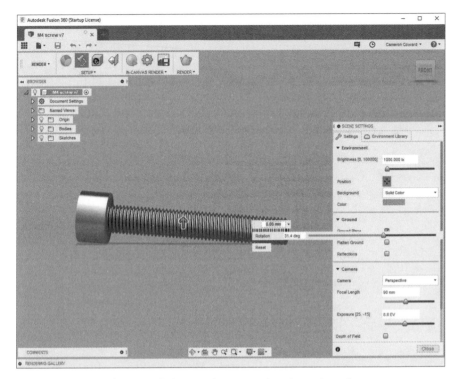

Figure 9-2: Position the ground plane so that it's just barely touching the screw.

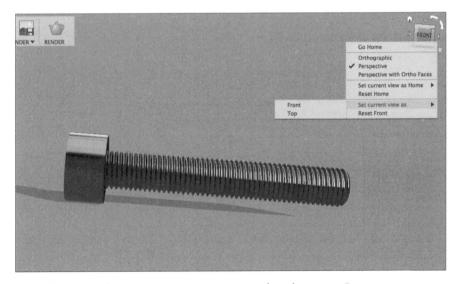

Figure 9-3: Rotate the screw to a resting position and set the view to Front.

After you've set the view as Front, the ground plane will rotate so it's parallel to the new bottom view. Your screw may end up floating up in the air again, but you can adjust the distance to the ground plane if necessary. When you're done, the screw should look like it's resting on a real physical surface.

Setting Up the Environment

Next you need to set up the environment. Changing the environment does a few things: it positions light sources, provides a background and ground plane, and determines what reflections to show on the surfaces of the model.

If you open Scene Settings again, you'll see a number of options for controlling how everything looks. Switch over to the **Environment Library** tab at the top of the dialog. There, you'll find built-in environments, along with a few that you can download from Autodesk.

Some of the outdoor scenes provided make the model look like it's in a real place, but most of the environments don't show a background behind the model in your rendered image. Instead, most, like Cool Light and Photobooth, are there to give your model realistic lighting, shadows, and reflections.

Later in this chapter, you'll learn how to place your model into a realistic scene, but for now, we'll focus on making the model itself look good. Choose the **Photobooth** environment and then switch back to the **Settings** tab. The first option is Brightness in Lux. Adjusting this value will increase or decrease the brightness of the light in the scene.

The Position setting, as we've already covered, changes the height and rotation of the ground plane. The Background drop-down menu lets you use either the environment background or a solid color. For this render, use the **Solid Color** option with the default gray. Below that, you can turn the ground plane and reflections on or off. The Flatten Ground option only has an effect if you're using an environment background, in which case it will make that background look less spherical.

Setting Up the Camera

The rest of the options in Scene Settings all relate to the settings of your virtual camera. Camera should almost always be set to Perspective, because this looks the most realistic. You could set it to Orthographic, which removes all lens distortion, but that's not how humans actually see things, so I recommend avoiding this setting. Focal Length has the same effect as it does on a real camera: a longer focal length has less distortion and flattens the image, while a shorter focal length makes the image look highly distorted (like a fisheye lens).

Also as with a real camera, you can adjust the Exposure setting to change how long your virtual camera's shutter remains open. This increases or decreases the overall brightness of the rendered image, instead of just altering the light sources with the Brightness setting. If you'd like, you can turn on Depth of Field to make the image lose focus the farther it gets from the center of focus.

The default settings are all fine here, but you can always adjust them if your rendered image is too dim or too bright. To prepare the image, close the Scene Settings and then click the **Render** button from the main toolbar and set Resolution to what you want your rendered image to be.

Performing the Render Operation

To finish, switch **Rendered With** to **Local Renderer**, which handles the computation on your own machine, instead of using the cloud-based Autodesk service. Autodesk provides a cloud renderer, but that requires credits to use. Use **Final** for the **Render Quality** setting and then click **Render**. Fusion 360 should then begin the rendering process, and a thumbnail should appear in the Rendering Gallery at the bottom of your screen. A progress bar should give you an approximation of how long the rendering process will take (in my case, it was about 30 seconds). Once it's done, you can double-click the thumbnail to open up your final render (see Figure 9-4). Then you can save or share it.

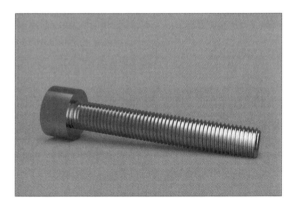

Figure 9-4: Your final render should look something like this.

This screw looks pretty good, but it's made up of a single part. Most of your models will be assemblies consisting of multiple parts, and the more detailed your model is, the better the final render will look. The screw, for example, would look more realistic if the head edges had tiny fillets.

Next, let's see how we can make a render look more realistic.

Rendering a Clock with a Decal Face

One of the best ways to make a model look real is to add decals, or 2D images. Almost every item you can purchase will have something printed on it, be it a logo, a warning diagram, or a button label.

To illustrate that concept, we'll model and render an analog clock using a decal for the numbers on the clock face.

Creating the Clock Parts

To start, model the physical parts of the clock to look like what you see in Figure 9-5. The clock should have a total of six components: the frame, the hour hand, the minute hand, the second hand, a pin for the hands, and a clear plastic cover. Create these however you'd like.

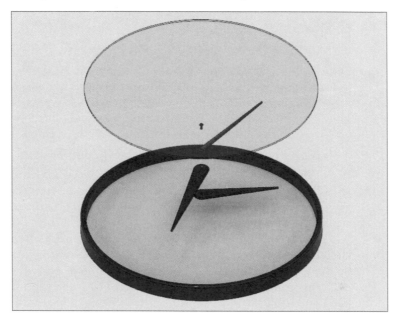

Figure 9-5: Model a clock with six separate components.

Next, assign a physical material for each part. The frame part will have two appearances: black plastic for the outer frame and white plastic for the face. After assigning the materials and appearances, use joints to lock them together. The hour, minute, and second hands should have cylindrical joints so that you can rotate them later to specific time settings.

Creating the Decal

Once you have modeled the parts, head over to Google and find an image of a clock face. Any image will work, but the higher the resolution, the better. That image will serve as the decal you'll place on the clock face, and pixelation from a low-resolution image will seem unrealistic. Ideally, you'll find an image with a transparent background to make it look like the graphics are printed directly onto the plastic face.

After you've found a suitable image, move over to the Render workspace to place it as a decal. Hide the clear plastic cover and click the clock face. Then, select the **Decal** tool from the main toolbar. For **Select Face**, choose the face of the clock, and for **Select Image**, choose the image you found. Turn off **Chain Faces** so that the decal isn't wrapped onto the black plastic of the frame. Then adjust the size and position of the decal, as shown in Figure 9-6.

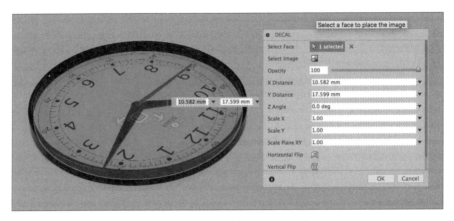

Figure 9-6: Select the face and the image and then position the decal.

Now that you've placed your decal on the clock face, you can set up the rest of the render. Move the hour, minute, and second hands so that they show the time you want; then show the clear plastic cover. Set **Grid Light** for the environment, because the harsh lighting will give the plastic cover some realistic light reflections.

Your Scene Settings should have **Ground Plane**, **Flatten Ground**, and **Reflections** checked. You can leave the camera settings at the default: 90 mm focal length and 9.5 EV exposure. Then just orient the model however you want it and create a final render. The result should look something like Figure 9-7.

Figure 9-7: The final clock render should look pretty nice!

As you can see, adding a decal adds realism to the detail of a 3D model render, as do reflections on the clear plastic cover of the clock. But the clock is still just sitting on a plain background—a dead giveaway that you're looking at a render.

Placing a Render in a Real Photo

To really sell the realism of a render, the best thing to do is place it in an actual photo. Ideally, you'll choose a photo of a setting in which you're likely to find your model.

To illustrate this concept, you'll learn how to place a rendered Arduino Uno microcontroller development board into a real photo of a desk. I downloaded the 3D model of the Arduino Uno from Autodesk's online gallery. Kevin Schneider created it, and it has a lot of detail. It even has labels on the capacitors. That detail is paramount to creating a photorealistic render.

Setting Up Your Background

To start, you'll need to take a photo. Arrange the desktop scene you'll place your model in. You'd likely use a real Arduino Uno with other electronics items, so for my background, I placed a soldering iron, breadboard, and other such items in the frame (shown in Figure 9-8). Before you take the photo, put some sort of placeholder on the desk to help you position the 3D render later on. I used a piece of paper that I cut to be slightly smaller than an actual Arduino Uno.

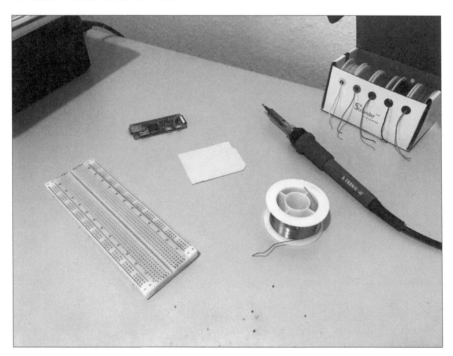

Figure 9-8: Set up your scene and add a placeholder for the Arduino Uno.

Once you've snapped the photo, bring your 3D model of the Uno into the Render workspace. If you downloaded the same model that I did, all of the detail (including decals) should already be there. Choose an Environment setting with lighting similar to the lighting in the room you took your photo in. Then use the Position Rotation angle to ensure that the light source is hitting the model at the same angle it would if it were actually in the photo.

Orienting Your Model

In the previous renders you created, the Camera settings weren't all that critical. In this case, however, we want to make sure the lens distortion in the render matches that of the photo you took. The most critical parameter is Focal Length. You can find that information in the metadata of your photo, or possibly from your camera's specs online if the focal length isn't adjustable. In my case, I took the picture on an iPhone 7, which has a 35 mm equivalent focal length of 28 mm. You can see my Scene Settings in Figure 9-9.

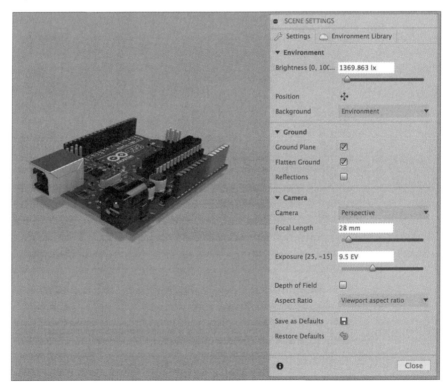

Figure 9-9: Change the Camera settings to match those of your actual camera.

Now orient the model to match the position it would have if it were really in the photo. Make the distance between the model and the camera the same as the distance in the photo you took. You need to match that distance both vertically and horizontally so the distortion of the render matches the distortion of the photo.

Once you've moved the model to the correct position, rotate it so that it's at the proper angle. That's where the placeholder you put in the photo comes in handy. It will help you tilt the model to the same orientation.

Getting the distance and angle just right can take a lot of trial and error, so for the next step, you may want to use low-quality renders until you've gotten the orientation right. You may need to create multiple renders, so making them lower quality will save time. When you create those renders, make sure you select the **Transparent Background** option.

Combining Your Images

The next step is to combine your rendered image with the photo you took. For that, I used GIMP (GNU Image Manipulation Program), which is a free and open source alternative to Adobe Photoshop. The particulars of using GIMP are outside the scope of this book, but you'll learn the basic process next. (It should be roughly the same in other photo editors.)

Open both the original photo you took and the new rendered image you just created. Then copy the rendered image to a new layer on top of the photo layer. Next, position and scale the rendered image so it's on top of your Arduino Uno placeholder. If the orientation doesn't look quite right, adjust your model and render a new image. Repeat that process until your rendered image exactly matches the distance and tilt of your placeholder; then switch to a high-quality render with a high resolution.

You might need to adjust the color of both the rendered image and the original photo so that they match, paying particular attention to the white balance. Because the render has a transparent background, it doesn't currently have a ground shadow. To add one, create a new layer under the render layer and over the photo layer and then use the Airbrush tool to paint a shadow on the new layer.

Once the rendered Arduino Uno looks like it's part of the original photo—shadows and all—flatten the image to merge the layers.

Finally, add some photographic noise, or graininess, to the image. Your original picture almost certainly had a small amount of noise, but the rendered image doesn't, which makes it stand out. A little bit of hue, saturation, and value (HSV) noise on top of the entire image will make it look more consistent. The final result should look like Figure 9-10.

Looks pretty decent, right? An image like this wouldn't hold up to photo forensic scrutiny, but it does a much better job of making your model look like a real object than if the model were on a blank background. Taking the time to place your renders in a scene will help you picture your models in the real world and go a long way toward selling others on your ideas.

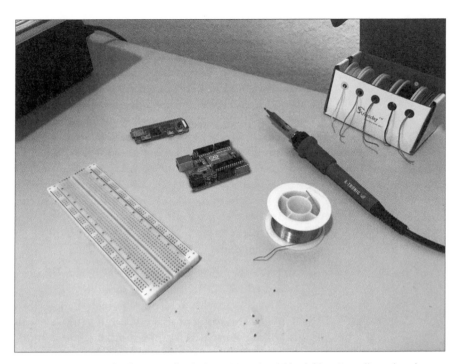

Figure 9-10: In your final image, the Arduino Uno should look like it's really in the photo.

Summary

When you started this book, creating a simple cube was probably exciting. Now, you've developed a valuable skill set you can use to design complex 3D models, make technical drawings to have them manufactured, and even create photorealistic renders of your creations. Whether you're using those skills to make 3D-printable replacement parts for your kitchen appliances or to design the next generation of high-tech robots, I hope you continue to practice what you've learned. But, before you go, I have one last exciting project for you in Chapter 10.

10

CAPSTONE PROJECT: CREATING A ROBOT ARM

For your final exercise, you'll show off the skills you learned in this book by designing a small robot arm. This arm, shown in Figure 10-1, uses a total of four small 9G hobby servos, which are motors you can rotate to specific positions with commands from a microcontroller development board. The completed robot arm will have a reach of about 7 inches. It won't be able to lift anything particularly heavy or operate with a great deal of precision, but the relatively simple construction is a good introduction to robotics and multipart assemblies.

This chapter won't guide you through every step of the modeling process. Instead, you'll learn enough to determine how to model the parts yourself. We also won't cover how to program this robot arm, as that's a complex topic in its own right. But, if you use a popular microcontroller development board like an Arduino, you can find many tutorials online that thoroughly explain how to control servos.

Figure 10-1: The completed robot arm

For this project, you'll need the following parts:

- Four 2.5 kg·cm stall torque 9G hobby servo motors (any brand will do)
- Arduino or similar microcontroller development board
- Sufficient 3D printer filament to print the parts, which should be less than 0.1 kg

This design doesn't use any bearings, and you'll only need the screws that come with your servo motors. A small amount of superglue will help hold the parts together, but you can also design a means of attaching them if you prefer.

Measuring and Modeling Your Servos

Although you'll have to buy the servos you'll use for this project, you'll also want to model them. While most 9G hobby servos are very similar in size and shape, there are often small differences between models from different manufacturers. To ensure that everything fits together, you need to model your servos exactly as they are.

Many manufacturers provide drawings with the servo dimensions, so you should first check on their website to see if those are available. If not, you can use a set of digital calipers to measure the servos yourself. Digital calipers can measure the distance between features far more precisely than rulers or tape measures. You can find inexpensive calipers online for less than $30. As you measure, pay particular attention to the areas where the servo will mate with other parts, like the screw mounts and motor shaft.

Modeling the Base of the Robot Arm

With the servos modeled, you can begin designing the 3D-printed parts of the arm. The first one—the base shown in red in Figure 10-2—is a teardrop shape that keeps the arm from tipping over when it's fully extended.

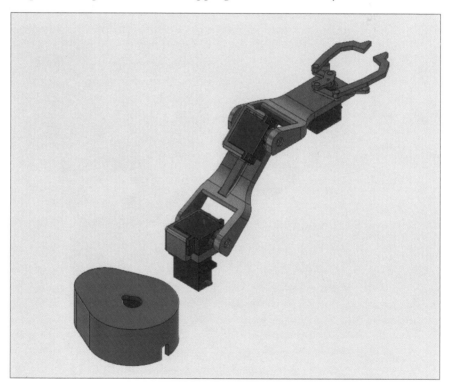

Figure 10-2: Model of the base, shown separated from the rest of the assembly

Start your design by creating a circle with a diameter of 60 mm; then elongate this to form the teardrop shape. The base should be wide enough to handle the load, but if your arm ever falls over, you can add additional elements to it for stability. The base should also be hollow. You'll fasten the first servo, which controls the rotation of the shoulder of your robot arm, to the inside of the base with screws through holes on the tabs that extend from the body of the servo.

The top of the servo body, where the shaft comes out, should sit a millimeter or two above the top surface of the base for clearance. Finally, place a small cutout at the bottom for the servo cable to exit through; that way, the base can still sit flat on the surface you place it on.

Once you've designed the base, put it in an assembly with the first servo motor.

Shoulder Motor Mount

The next part you'll model is the mount for the second servo motor, shown in Figure 10-3. This part will connect the first servo in the base to the second servo. The first servo rotates the entire arm relative to the base. The second servo tilts the arm up and down. The mount should cradle the second servo so it will lie on its side.

Figure 10-3: The second servo mount joins servos 1 and 2.

Create a hole in the center of the mount's bottom for the first servo's shaft to go through. The hole should also have a cavity on the top to keep the shaft screw's head below the surface. Next, create two more small holes for the servo-mounting screws; align these with the holes on the servo. Finally, create a channel and hole to allow the servo cable to pass through the mount. Pay attention to the second servo's orientation in Figure 10-3 and make sure your channel and hole are on the side where the servo cable will exit the mount.

Create another hole in the mount opposite the servo's output shaft but directly along the shaft's axis. You'll use this hole for a pin that acts as a pivot point for the next segment. It must be on the same axis as the servo motor shaft to ensure smooth motion. Once you've modeled the mount, add it to your assembly.

Third Motor Mount and First Segment

This next part, shown in Figure 10-4, will act as both the first segment of the arm and as a mount for the third servo. It has parallel pivot points on each end—one for the second servo and one for the third servo. The pivot points are, on one side, the pin and hole, and, on the other side, the servo hub shaft and pin hole. You can reuse some of the geometry from the shoulder motor mount as a starting point, because you'll mount the third servo just like the second servo.

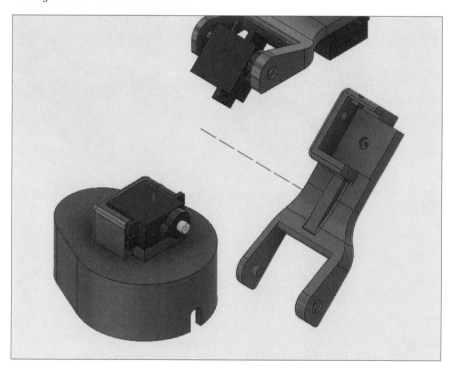

Figure 10-4: The third servo mount is also the first segment of the arm.

The distance from the second servo pivot axis to the third servo pivot axis should be about 70 mm. Give the other side a U shape so that it can fit around the second servo and mount. Create a hole on one side to fit over the servo shaft and a pin on the opposite side to fit into the hole on the mount. It's a good idea to add a rib (the vertical bar in the middle of the part) to give the part some rigidity. Add it to the assembly model and make sure it can rotate unobstructed.

Fourth Motor Mount and Second Segment

As with the first segment model, you can start the second segment model by reusing some geometry. The area where the second segment mounts to the third servo is exactly the same as where the first segment mounts to the second servo. As you can see in Figure 10-5, the section you have to create from scratch is the other end, which attaches to the gripper mechanism.

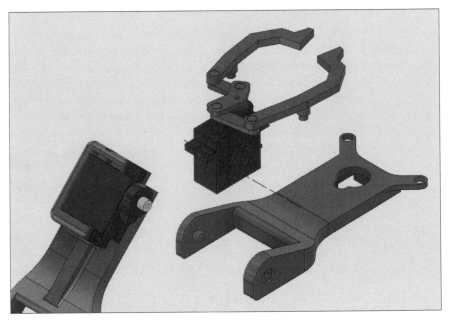

Figure 10-5: Reuse the side of the second segment that's the same as the first segment.

On the gripper mechanism side, create a hole, like the one you created on the top of the base, for mounting the fourth servo. At the end of that side, model two 45-degree protrusions with 4 mm diameter holes that will be the pivot points for the gripper arms. You should make each protrusion roughly 12 mm long. Once again, add the part to the assembly to make sure it can move freely.

The Gripper Mechanism

The robot arm's gripper mechanism is the most complex piece of the entire project. A hub on the motor shaft connects to two short linkages, which in turn connect to the arms of the gripper. When the hub rotates, it pushes the linkages out, causing the arms to pivot and close.

The mechanism is complex because the lengths of servo hub arms, the linkages, and the "fingers" of the gripper will affect how well it operates. Get one of these items really wrong, and the entire gripper will jam up. For that reason, you'll want to spend some time experimenting with different lengths in CAD before you actually 3D print any parts. Start with lengths that look similar to Figure 10-6 and add joints for them. Then, move the mechanism to see if it opens and closes fully. If it doesn't, adjust the lengths and try again.

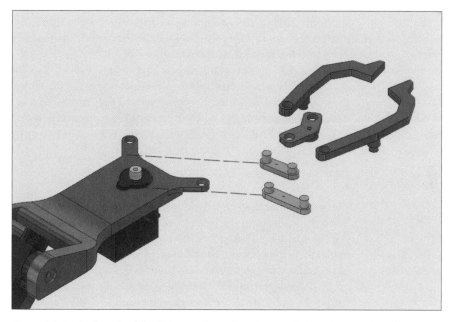

Figure 10-6: The length of the linkages determines how well the gripper will work.

When you're modeling these, you'll need to make one arm longer to avoid jamming. To compensate for this asymmetry, you can make one side of the center hub longer than the other. This allows the gripper to open by roughly equal amounts on each side.

Finally, take some care with the linkage pins and the holes they fit into. The pins need to have a large enough diameter to be strong, so they won't break, and the fit needs to be loose enough to allow free movement. You can 3D print small discs to superglue onto the pins after assembly to hold the linkages in place, or you can design the pins with tabs so they snap into place.

Add the parts to your assembly when you've finished modeling them and make sure they all fit together.

Printing the Parts and Assembling the Robot Arm

You can print all of these parts in PLA or ABS on any hobby 3D printer, including a fused-filament fabrication (FFF) printer. Only the second segment will require support material during 3D printing. You'll want to keep the parts' density high enough to be strong but low enough to be lightweight. Somewhere between 25 percent and 50 percent infill density should work well.

Once you have all of your parts, assembling them is as easy as mounting the servos with the provided screws and snapping the pieces together. The second and third servo pivot mounts will fit tightly, but a little force should get them on. After you've placed each part onto the corresponding servo motor, use the included shaft screws to hold them on securely. Then just place a small dab of superglue on the gripper linkage pins to attach the discs and keep the pins from sliding out.

With the arm assembled, you can attach the servo cables to your microcontroller development board to control the arm. These 9G hobby servos use only a small amount of power, so you may not need a separate power supply; check the specs of your motors to see how much current they use and what your microcontroller development board can supply.

Summary

Before you started reading this book, you probably didn't know how to begin modeling an entire robotic arm. Now, with only minimal guidance, you should have done just that by drawing on what you've learned. The project in this chapter shows just how much you can accomplish with the CAD skills you learned in this book.

You can continue to develop these skills as you go forward, whether for hobby projects or in your professional life. While this book didn't cover some of Fusion 360's most specialized tools, your working knowledge of the software and CAD modeling practices should allow you to understand them if you ever need them. By this point, you should understand *how* parametric CAD works. Getting good at 3D modeling requires more than simply knowing how to use a tool. It takes a specific kind of thinking and forethought. Now, take that thinking and apply it to your next project!

INDEX

A Beginner's Guide to 3D Modeling is set in New Baskerville, Futura, and Dogma. This book was printed and bound at Versa Printing in East Peoria, Illinois. The paper is 70# Skyland Offset.

The book uses a layflat binding, in which the pages are bound together with a cold-set, flexible glue and the first and last pages of the resulting book block are attached to the cover. The cover is not actually glued to the book's spine, and when open, the book lies flat and the spine doesn't crack.